FOUNDATIONS
OF
FLYING

To James A. Marler, a man of the sky.

FOUNDATIONS
OF
FLYING

BY JEFF W. GRIFFIN

TAB BOOKS Inc.
BLUE RIDGE SUMMIT PA 17214

FIRST EDITION

FIRST PRINTING

Copyright © 1982 by TAB BOOKS Inc.

Printed in the United States of America

Library of Congress Cataloging in Publication Data

Griffin, Jeff.
    Foundations of flying.

    Includes index.
    1. Airplanes—Piloting. 2. Flight training.
I. Title.
TL710.G77   1982   629.132′521   82-5858
ISBN 0-8306-2345-0 (pbk.)   AACR2

Contents

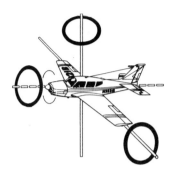

Introduction

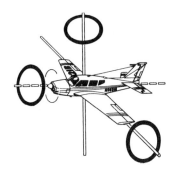

In any field of endeavor there are several sources of information. In aviation we are lucky to have such a wide range of literature to choose from. Before people go into a hobby or career, they should learn everything they possibly can about that subject. In this way they can save themselves tons of money and hours of time. If they choose not to continue their interest, they will have gotten out before spending any more money than was necessary to buy the literature. If they continue to have an avid interest in the subject, they will soon learn which is the best way to go about getting their feet wet. And this will also save them money and time. This book fits into either category.

Within the succeeding pages is several years worth of the instructor's point of view about learning to fly. All the basics are here that concern the manipulation of the airplane. Many basic primers on flying try to encompass everything that is needed to know about passing the first Private Pilot exam. This book contributes quite a bit to that idea. Yet, there is no chapter on navigation or engine performance or weather. That information can be found in a multitude of books. What the aspiring pilot will find within these pages is pure flying know-how. In other words, how to manipulate the airplane and get the results desired.

I can proudly point to my instructional record. Never has one of my students ever failed his check ride on the first try. Although I no longer instruct, I live by what I preach every time I show up for work. This book incorporates all the little learning tricks that were taught to me by other instructors and those that I have learned.

Nothing is held back. If I think a maneuver may cause the average pilot a little apprehension, I talk about it. When I learned to fly there were lots of little things that scared me. It's only natural, and as a result it has made me a safer pilot. It will do the same for you.

In several places in this book I break a maneuver into a simple 1-2-3 method. I had a great deal of success with this approach because it allows the student to concentrate on one portion of the maneuver at a time. The upshot of this is a higher proficiency in less time than with other techniques. This book should be of great value to you as you learn or prepare to learn.

This book probably has the greatest value as a prelesson text. If your instructor outlines the next lesson for you ahead of time, this book will indeed enhance your performance. Inside these pages are the common errors that all of us tend to make in the learning-to-fly process. If we know what they are ahead of time, we can prepare for them. Study these, and chances are you will stun your instructor with your ability to pick a maneuver apart the first time it is introduced.

Chapter 1

The Wing and Principles of Flight

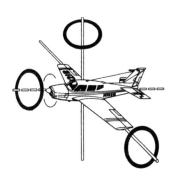

Have you ever compared wings on aircraft? If you have, you have probably noticed that they can vary a great deal. For example, the wing of a Cessna 152 is big and fat. The wing on a T-38 Air Force jet trainer is almost razor-thin by comparison. However, a very close inspection will reveal that they are just altered versions of the same thing. The basic wing (or *airfoil*, as some like to call it) is designed to do one thing: provide lift. The reason that wing designs vary from one aircraft to the next (and we are talking about types of airplanes) is that a change in design may provide better performance in some area, such as climbing performance. Another design may give better low-speed performance so that the airplane may take off and land at slower speeds and use less runway. So wings are designed for the purpose the airplane is to serve. Regardless of that purpose, though, all wings share common design criteria and nomenclature. In the most basic of terms, an airfoil is curved on the upper portion and nearly flat on the bottom. This plays a significant part in producing lift as we will see.

The Four Forces

The day I took my first flying lesson was quite a day. The magic of the little trainers buzzing by as they did their touch-and-goes filled me with expectation that I would soon be up there enjoying the thrill of flying with them. Not only did I begin to learn to fly that day, but aviation crept into my blood and forever changed the direction of my life. Possibly, the difference between myself and others who aspire to fly was that the foundation that was laid down for me was solid.

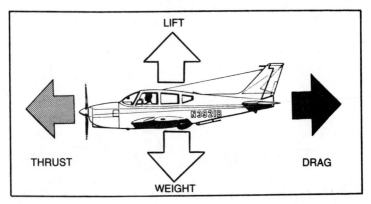
Fig. 1-1. The relationship of forces in flight.

My instructor began that day with the "four forces." These act upon an airplane during straight and level unaccelerated flight (Fig. 1-1). The upward-acting force is *lift*; the downward-acting force is *thrust*; and, finally, the backward-acting force is *drag*. Simply, lift opposes weight and thrust opposes drag. In straight and level flight, where the airspeed remains constant, all these forces are equal to each other. Drag and weight are forces that are inherent in anything lifted from the earth and moved through the air. On the other hand, thrust and lift are created artificially to overcome the other two forces supplied by nature. The engine/propeller combination produces the thrust and the wing produces the lift.

The importance of the balance of these four forces is this: lift equals weight and thrust equals drag. However, lift and weight will not equal thrust and drag. If any inequality occurs between lift and weight the airplane will begin a climb or descent. The same holds true for drag and thrust. If they fall out of balance, acceleration or deceleration will occur. In either case, the induced condition will continue until the forces once again come into balance.

Relative Wind

When discussing the principles of flight it is essential that all the terms are understood. One of the more difficult terms to understand is *relative wind*.

Hold your hand up wherever you are sitting. Do you feel any wind acting upon your hand? You probably feel little or no wind depending on whether you are sitting inside or outside. But how about if you are riding in a car and put your hand out the window? You feel a strong action on our hand by the wind. Relative wind, then is the force we felt as an object is moved through the atmosphere. It is not to be confused with the natural wind. To illustrate further, if you are riding in the car with your hand out the window, can you tell which way the natural wind is blowing by the action on your hand? The answer is no. The natural wind has no action on the relative wind.

So what has all this to do with an airplane flying through the air? Well, an airplane moving straight ahead through the air experiences relative wind just as the car does. If you put your hand out the window of an airplane (and I don't advise it) the effect will be the same as in a car. The relative wind works in the opposing direction to the flight path of the airplane (Fig. 1-2). Regardless of the aircraft's attitude (climb, straight and level, or descent), the relative wind's direction remains unchanged.

I once had a student who really did not have a grasp of either relative wind or natural wind. I somehow had never uncovered this fact until I accompanied her on a cross-country flight. While we were airborne, I noticed that we were beginning to drift off course. Since her preflight planning was correct, I assumed that she had forgotten which side the natural wind was from and was not correcting for it. In order to bring this to her attention, I asked her if she knew which direction the wind was coming from.

"Sure," she said. "It's coming from straight ahead of us." A little surprised at her answer and knowing that it was incorrect from her own preflight planning, I queried further.

"How do you know where the wind is coming from?" I asked.

"Well, look at the clouds," she said. "They're coming right at us!"

"Indeed they are," I conceded, "and that might tell the story if we were standing on the ground looking up. But what if you are in a car on the highway. Don't the sign posts and telephone poles on the side of the road also move towards you the same way the clouds are now?"

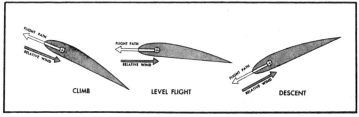

Fig. 1-2. Relationship between flight path and relative wind.

"Oh, I see what you mean now!" Then, after a few seconds: "but then where is the wind coming from?"

"Look at your preflight planning," I said, "and you'll know."

Indeed, the clouds were moving directly towards us as my student so aptly observed. The trouble with her answer was that she was pointing to the relative wind and not the natural wind. (Actually, *we* were moving toward the clouds.) Remember, the relative wind is always straight ahead whether we are climbing, flying straight and level, or descending.

Angle of Incidence

It is not widely understood by most pilots flying today that the wing is not mounted to the body with the chord line parallel to the longitudinal axis of the aircraft. Instead, it is usually mounted at an angle such as the one shown in Fig. 1-3. This angle is known as the *angle of incidence.* The main reason for mounting the wing in this fashion, incidentally, is to make the airplane easier to control during landing and takeoff. By easier, I refer to control pressures exerted on the controls in the cockpit.

How the Wing Produces Lift

The wing is probably one of the greatest inventions since the wheel. Although it is well understood by engineers of today, it was not well understood by pioneers such as the Wrights or Lillenthal. If it had been, they would have made progress a great deal sooner. Essentially what the wing or airfoil does is to create a low pressure area on the top surface of the structure and a high pressure area on the bottom side of the structure. The difference between those two forces is the *lift.* The high pressure below the wing tries to migrate to the area of low pressure and the wing is in the middle. As a result, it is lifted.

In Wolfgang Langwiesche's book *Stick and Rudder,* he states that the wing actually pushes air down. If one flew only an inch

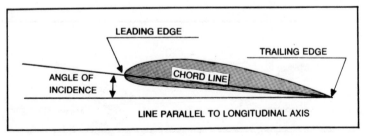

Fig. 1-3. Cross-sectional view of an airfoil.

above a cloud that cloud would be pushed down as a result of the passing wing. Mr. Langwiesche also urged that Bernoulli's Principle not be the focus of a pilot's understanding about why the airfoil produces lift. While the wing does indeed exert a downward force as Mr. Langwiesche describes, it is this author's opinion that the scientific principle set forth by Bernoulli is of the utmost importance in understanding how the wing reacts as it passes through an air mass. The following is my interpretation of how that principle works.

Let's say that we are plumbers for a minute. We have constructed a pipe that is four inches in diameter and choked down to a two-inch diameter pipe and then back up (in size) to a four-inch diameter pipe. It will look just like the one in Fig. 1-4. Now let's attach a pressure gauge at each of the diameters of pipe. Each four-inch portion will have a gauge and the two-inch portion will have one also. Now, if we pump water through this pipe system we will notice that both four-inch gauges are indicating the same pressure. This means then, that both sections of the four-inch pipe are full of the same volume of water. The amazing thing about this arrangement is that the ends hold the same pressure with a half-sized pipe in between them.

What do you think is happening? In order for both the larger pipes to be full at the same pressure, the smaller pipe must carry water through to the next larger pipe at a higher velocity. This keeps all the pipes full, but a very important natural occurrence is taking place. Since the water in the smaller two-inch pipe is having to move at a higher velocity the pressure is reduced. The center

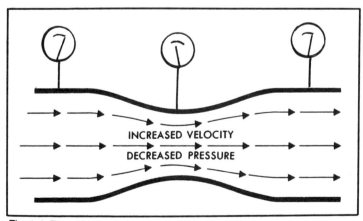

INCREASED VELOCITY

DECREASED PRESSURE

Fig. 1-4. Flow of air through a Venturi tube. Note the pressure gauges.

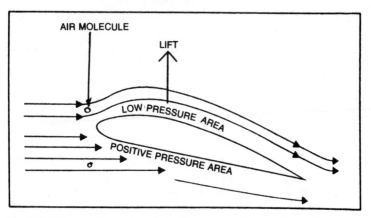

Fig. 1-5. Difference in pressure between upper and lower wing surfaces produces lift. The air molecules will meet at the same time, but the top one must travel farther.

gauge in the illustration will indicate appropriately low. This is a most important point. Since the atmosphere behaves as a fluid with respect to aircraft, we can expect air moving at a higher velocity than surrounding air to have a lower pressure.

Now I know that the relationship between plumbing and flying is a little far-fetched. But there is a connection which we will discuss. Note the shape of the airfoil in Fig. 1-5. There is much more wing above the chord line than below it. Let's extend the chord line out in front of the wing. If we put an air molecule on each side of this line and move the airfoil forward, one molecule will travel above the surface of the airfoil and one will go below. Since the upper surface of the airfoil is more curved than the lower surface, the air molecule on the upper portion of the wing will have to travel a little farther in about the same time to meet its partner which is traveling below the wing. To travel farther in the same amount of time requires the top molecule to travel faster. Therefore, if one molecule travels faster, then all molecules on the top must travel just as fast. This causes a lower pressure on the top of the wing. Remember, the atmosphere behaves as a liquid or fluid with respect to an aircraft. Thus, the connection between the plumbing and the wing is that fluids moving faster exert less pressure.

Now, if you haven't got a clear picture of what is happening, look at Fig. 1-6. The faster moving air over the top of the wing is producing a lower pressure with respect to the higher pressure beneath the wing. It is this higher pressure below the airfoil that is acting to push the wing up—and the rest of the airplane with it.

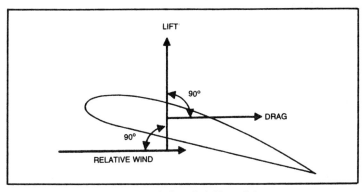

Fig. 1-6. Relationship between relative wind, lift, and drag.

Angle of Attack

The angle of attack is the angle between the wing chord line and the direction of the relative wind, or the flight path of the wing. This angle should not be confused with the angle of incidence. Remember, the angle of incidence has to do with how the wing is mounted to the fuselage of the aircraft. The angle of incidence is constant. The angle of attack on the other hand, changes from moment to moment during flight. Look at Figs. 1-7 and 1-8. They show how the angle of attack can change for the same flight path or remain the same for different flight paths.

Angle of attack plays an important part in the amount of lift any given airfoil can produce. At zero angle of attack, the pressure below the wing is equal to the atmospheric pressure. This is because the relative wind is passing straight by the wing on the bottom side without being deflected in any manner. In this case, all of the lift would be produced by the decrease in pressure (less than atmospheric pressure) along the upper surface of the wing. At small angles of attack the *impact pressure* (or *positive pressure*, as it is sometimes called) is negligible. Just to make things clear, there is some amount of lift generated strictly by wind hitting the upturned

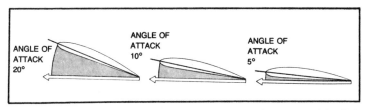

Fig. 1-7. The angle of attack is the angle between the wing chord and the flight path.

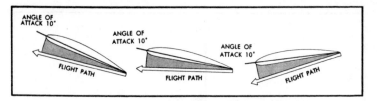

Fig. 1-8. The angle of attack is always based on the flight path.

bottom of a wing. This is called impact pressure. In cases, though, where the angle of attack is zero or near zero, most of the lift is produced in accordance with Bernoulli's theorem.

As the angle of attack is increased, the impact or positive pressure on the bottom surface of the airfoil will increase. Also, the pressure above the wing will continue to decrease following the principle. The pressure above the wing continues to decrease — that is, as long as the air flows smoothly and follows the curvature of the wing. Of course, the curvature of the wing is effectively increased as the angle of attack is increased. The combination of the increased *camber* (curvature) of the wing and the increased impact pressure increases the upward force or lift greatly. It also coincidentally results in greater drag.

When the angle of attack increases to about 18° to 20° on most airfoils an important condition occurs. The air can no longer flow smoothly over the top of the airfoil or wing's upper surface. It is forced to flow straight back from the leading edge of the wing and away from the top surface of the wing. This causes a certain amount of eddying as in a swift trout stream behind a rock. This is also known as burbling and there is an angle named for it. Where burbling begins the angle is known as the *burble point*. If the angle of attack is increased further at this point the burbling which began at the trailing edge of the wing will very quickly work its way forward towards the leading edge. This results in a sudden increase of pressure on the top of the wing. As one can imagine, that leads to a great loss of lift and an increase in drag. This condition is known as an *aerodynamic stall*. The remedy is to decrease the angle of attack. This sequence of events is shown quite well in Fig. 1-9.

Thrust and Drag in Straight and Level Flight

As we mentioned in the section on the "Four Forces," thrust works against drag and vice versa. Drag is the hardest to understand. Thrust is simply generated by the engine and transmitted into useful energy by the propeller.

8

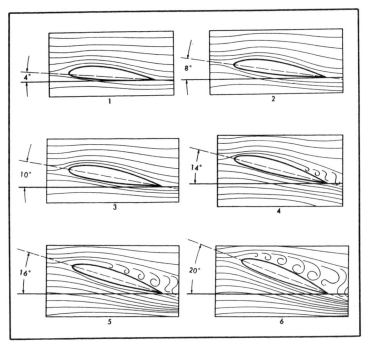

Fig. 1-9. Flow of air over a wing at various angles of attack.

Drag, at first, can seem mysterious. This can be attributed to the different name we give the item in aviation instead of the usual name of "friction." *Drag is in effect nothing but friction.* When we rub our hands together they become warmer, because the rubbing of the hands together results in friction. It works the same way with a wing. The air flowing across the wing causes friction or drag. Even though most airfoils are extremely smooth, they cannot avoid friction or drag between the skin of the wing and the air moving past it. The faster the air moves by, the more friction or drag is generated. Within that fact is the relationship between thrust and drag.

If we are flying straight and level and increase the power of the engine by pushing in the throttle, the thrust is increased. If we continue to hold the aircraft in a straight and level attitude, the speed must necessarily increase. As the speed increases, so does the drag. At some point, the amount of power that we have added will not be enough to overcome the increase in drag. The result is that the airplane will no longer accelerate. A straight and level, unaccelerated condition will occur as before, at the new, higher level of power or thrust.

Whenever thrust is decreased, the relationship between thrust and drag is still in effect. If we reach down and pull the throttle back, power to the propeller is decreased and so is thrust delivered by the propeller. If we endeavor to maintain straight and level flight, the speed of the aircraft through the air will decrease. As it does, the amount of drag will also decrease. When the two forces equal each other, a constant airspeed (speed through the air) will result.

Lift and Gravity in Straight and Level Flight

Lift, as we have already discussed, is the upward acting force on an airfoil. Lift always acts perpendicular to the relative wind. Lift is the force that counteracts the weight of the airplane. Whenever lift is equal to the weight of the plane, the aircraft will neither gain or lose altitude. Of course if the lift is decreased and weight is greater than lift, the aircraft will descend.

Factors Affecting Lift and Drag

There are so many variables involved in the flight of an aircraft that it has been compared to balancing a bowling ball on a ball bearing. In fact, there are many changes going on constantly during each and every flight just to make the airplane fly as it should. The items that affect lift and drag are wing area, shape of the wing, angle of attack, velocity of the relative wind (airspeed), and density of the air moving over the wing. A change in any of these items affects lift and drag or the relationship between lift and drag. Each means of increasing lift will also increase drag.

Effect of Wing Area on Lift and Drag. The lift and drag acting on a wing is roughly proportional to the area of the wing. If we double the area of a wing, the lift and drag on that wing will also double. The only way we as pilots can change the surface area of wing in flight is by the use of certain types of flaps. One of the most effective type is the Fowler flap. This type of flap moves backwards away from the trailing edge of the wing at the same time it moves in a downward direction. This increases the wing area and thus increases lift as well as drag.

Perhaps we should discuss one other type of drag here—*form* drag. Form drag is that drag induced by the form of the airplane as if it were two dimensional. For example, stand in front of an airplane (or a car, for that matter). If we position ourselves directly in front of the vehicle, we can only see the front of the object. We cannot see the sides. The portion of the vehicle that we do see must be forced directly into the relative wind. The form of the vehicle induces drag.

How this relates to increased wing area is simple. When Fowler flaps are lowered, we can see them lowering into the position of the relative wind. In effect, not only has the skin friction been increased by increasing the area of the wing, but form drag has been increased as well.

Effect of Airfoil Shape on Lift and Drag. Earlier in this chapter, we discussed how airfoil design can increase lift characteristics. The upper curvature of an airfoil (camber) can be increased up to a certain point to increase or enhance lift characteristics. High-lift wings have a large curvature on the upper surface with a lower concave surface. Wing flaps can be used to approximate this condition (Fig. 1-10). As you can see, lowering a flap arches the upper surface and concaves the bottom. The same sort of effect can be achieved by lowering an aileron. If the aileron is lifted, however, the upper surface is shortened and the area is decreased. Lift decreases. The elevator accomplishes for the horizontal stabilizer the same sort of thing. It also changes the direction of lift. The rudder does the same thing on the vertical tail surfaces. Of course, they all work in concert with each other.

The shape of an airfoil can also be altered by the accumulation of ice on the leading edge. Many people believe that ice forming on an airfoil causes the plane to descend because of the weight. The amount of weight added to a plane in flight in this manner is negligible. What in fact does happen is a disruption of airflow over the top surface of the wing. Although the surface of the wing is increased by the accretion of ice, the ice is never aerodynamically correct. The airflow is disrupted and interferes with the highest lift-to-drag ratio designed into the wing. The result is a loss of lift and an increase in drag.

Even the slightest coating of frost on a wing can prevent an airplane from producing enough lift to take off. The smooth airflow over the wing is disrupted and the lifting capability of the wing is destroyed. The only thing to do is to remove all frost, snow, and ice before takeoff is attempted. The failure to do this could result in a quick landlubbing ride through the weeds.

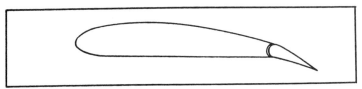

Fig. 1-10. Use of flaps increases lift and drag.

Effect of Airspeed on Lift and Drag. An increase in the velocity of the air passing over the surface of a wing increases the lift and drag. The lift is increased because of a higher impact pressure on the bottom surface of the wing as well as from the higher speed across the upper surface of the wing (Bernoulli's Theorem). Drag is also increased because in any increase in lift corresponds to an equal increase in drag (Fig. 1-11).

Wind tunnel tests have shown that lift and drag vary as the square of the velocity. This means that if the airspeed is doubled, the lift and drag will be quadrupled. This assumes that the angle of attack remains the same.

Effect of Air Density on Lift and Drag. In general aviation literature, there seems to be one subject that pops up again and again—*density altitude*. We won't discuss density altitude in specifics at this point, but rather we'll look at how air density affects lift and drag. It is essentially the same thing; only the numbers in

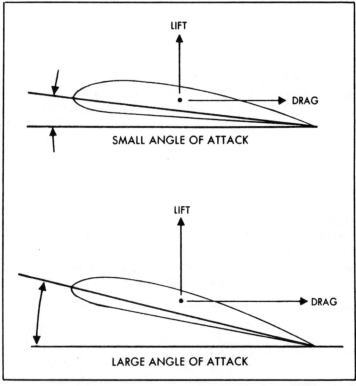

Fig. 1-11. Lift and drag increases with an increase in angle of attack.

calculating density altitude are not mentioned here. So what is density altitude? It is the varying density of the air and how it increases or decreases performance of a wing. And why does it pop up so much in aviation literature? Well, density altitude, on the negative side of the coin, can rob a plane of its ability to fly at warm temperatures and that results in accidents. Therefore, many magazines and books point out the dangers of ignorance on the subject.

If one thinks about it, it is easy to understand that the air becomes less dense when it is heated. The air also becomes thicker when the temperature decreases. This varying density affects the performance of an airfoil. If air density decreases, then lift and drag decrease. If air density increases, then lift and drag increase. Several factors affect the density of the atmosphere. Pressure, temperature, and humidity all act in concert to change the value of air density (Fig. 1-12). At an altitude of 18,000 feet, density of the air is half that at sea level. If an airplane is to maintain lift, the velocity of the air moving over the wing must be increased and/or the angle of attack must be increased. Because of this condition, airplanes require a longer takeoff run at higher altitudes than under the same conditions at lower altitudes. For example, an airplane taking off at an airport in Houston, Texas, is very near sea level. The takeoff distance on a day when the temperature is 80° might be 3500 feet. If the pilot flies to Denver, Colorado, and lands and takes off again and the temperature is also 80° there, the takeoff distance might be 6500 feet. These are not accurate figures, but are indicative of what happens as air density is decreased.

It is also true that an airplane can use more or less of the same runway on different days. For example, our same airplane and pilot in the previous example depart Denver on a day when the temperature is near zero degrees. In this instance, the airplane may only need 4000 feet of runway, 2500 feet less than it used on a warm summer day.

Humidity plays an important role as well in the density of the air. Water vapor weighs less than an equal amount of dry air. This condition is known as *high relative humidity* when the air is moist. When the air is dry, the relative humidity is low. To make it perfectly clear, moist air is less dense than dry air. Therefore, whenever other conditions remain the same, a high-humidity day will cause the takeoff distance of an aircraft to be longer. This situation can be amplified and especially ticklish when the temperature is high along with high relative humidity. Warm air can hold

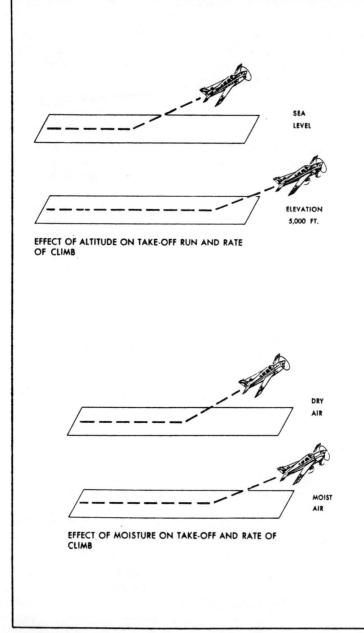

EFFECT OF ALTITUDE ON TAKE-OFF RUN AND RATE OF CLIMB

EFFECT OF MOISTURE ON TAKE-OFF AND RATE OF CLIMB

Fig. 1-12. Effect of altitude, temperature, and humidity on takeoff run and rate of climb.

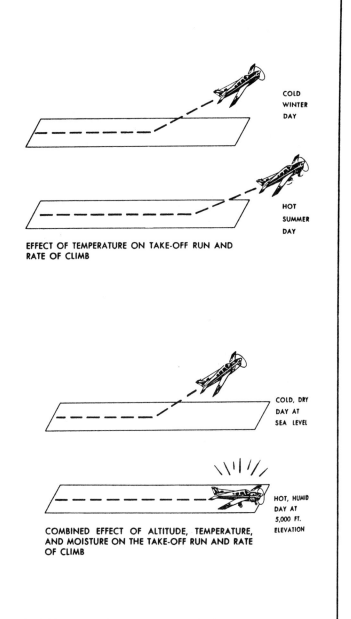

COLD
WINTER
DAY

HOT
SUMMER
DAY

EFFECT OF TEMPERATURE ON TAKE-OFF RUN AND
RATE OF CLIMB

COLD, DRY
DAY AT
SEA LEVEL

HOT, HUMID
DAY AT
5,000 FT.
ELEVATION

COMBINED EFFECT OF ALTITUDE, TEMPERATURE,
AND MOISTURE ON THE TAKE-OFF RUN AND RATE
OF CLIMB

15

more water vapor and remember, water is less dense than dry air. The problem is obviously compounded and takeoffs can become extremely long as compared to an average day.

We have already discussed how less dense air affects the airfoil. But that's not all that is affected by the density of the air. When the air becomes less dense, engine horsepower falls off. The propeller also loses some of its efficiency because the power output of the engine has been decreased and, because the prop is an airfoil itself. This all translates into longer takeoff runs. The propeller can't take as big a bite of the air under these conditions and thus the airplane accelerates slowly. This, then, causes a delay in the main airfoils becoming effective. Upon climbout, the rate of climb will be noticeably less for the same reasons.

From this discussion it should be apparent that a pilot must be on his guard for prolonged takeoffs in hot and humid weather. Remember that not only the temperature and humidity but the airfield *elevation* is extremely important. It is common in the summer months for the operating altitude of the airplane (density altitude) to be twice as high as the airfield elevation or the altitude at which one is flying.

Function of the Controls

By the time you have decided to pick up this book and read it, you are probably already familiar with the names of the control surfaces on an airplane. They are, of course, the *ailerons, elevator* and *rudder.* These control surfaces are used to control the aircraft about three axes of movement. In Fig. 1-13 we can see those three axes. We can think of these axes as axles about which the airplane can turn like a wheel. The *longitudinal* axis runs from nose to tail. This is the axis of *roll.* Movement about this axis is produced by the ailerons, which are movable surfaces on the trailing edges of the wings.

The *lateral* axis is the axis of *pitch.* This axis runs from wingtip to wingtip. Control of the aircraft along this axis is done by the movement of the elevator. This is what raises and lowers the nose of the aircraft in order that it may climb and descend.

The *vertical* axis runs from the top to the bottom of the airplane and intersects the other two axes where they intersect. Movement about this axis is not often noticed since it is usually coordinated with movement of the longitudinal axis. Singular movement of the vertical axis results in an uncomfortable *yawing* motion where the nose of the aircraft swings left and right. Movement about the vertical axis is controlled by the rudder (Fig. 1-14).

Control Surfaces

In a conventional aircraft, there are two ailerons, one on each wing near the end. They work in opposite directions to each other. When the left aileron is deflected down, the right is deflected up. The aileron that is deflected down increases the curvature of the top of the wing and increases lift. This wing will go up. The wing with the aileron deflected up will have decreased lift due to the upper wing surface being shortened. This wing will go down. The ailerons are moved simultaneously; when one is up the other is down.

The ailerons are interconnected by rods or cables and pulleys to the control yoke or stick. When pressure is applied to the right on the control yoke, the left goes down and the right aileron goes up. This begins a roll to the right. The aerodynamics are as we just discussed. The aileron that is raised shortens the camber of the

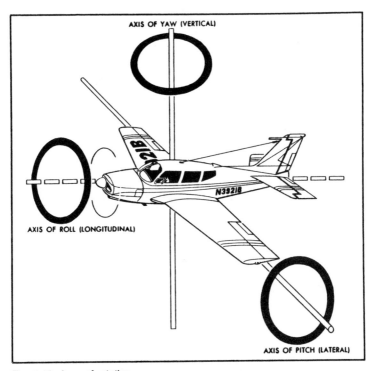

Fig. 1-13. Axes of rotation.

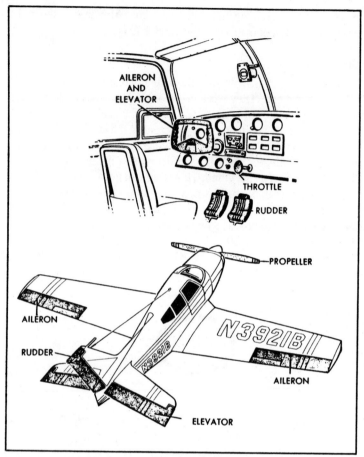

Fig. 1-14. Conventional arrangement of controls (shaded) is shown at top. The airplane surfaces that respond to the controls (shaded) are at bottom.

wing and decreases lift as well as the angle of attack which also reduces lift (Fig. 1-15).

In the airplane, during a preflight runup, most instructors insist that the student check the flight controls. The checklists in most planes also call for this item and the response to it is "free and correct." If you are like me, it is very hard to remember which aileron is supposed to go up and which is to go down with respect to the control yoke or stick. There is a crutch however, to help us dummies remember on the spur of the moment. Usually, the student will fly an aircraft equipped with a control wheel or yoke with his left hand *only*, leaving the right hand free to make adjustments with the throttle and flaps. Imagine your left hand holding on to the left horn of the yoke. Now, raise your thumb, pointing it towards the roof. If we apply pressure on the yoke to the right, the thumb will be pointing towards the right side of the aircraft. Remember this: the thumb always points towards the *up* aileron. Look out through the window and you will see that the aileron is indeed up. It works the same way to the left. If the control yoke is moved towards the left the thumb will point towards the left aileron, which should be raised.

In aircraft equipped with a joystick, this trick is not as apparent. The stick usually only moves about 40° from vertical at most.

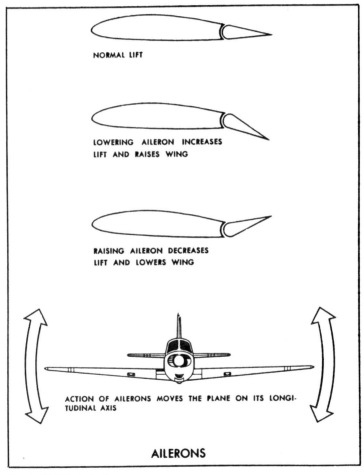

NORMAL LIFT

LOWERING AILERON INCREASES
LIFT AND RAISES WING

RAISING AILERON DECREASES
LIFT AND LOWERS WING

ACTION OF AILERONS MOVES THE PLANE ON ITS LONGI-
TUDINAL AXIS

AILERONS

Fig. 1-15. Effect of ailerons.

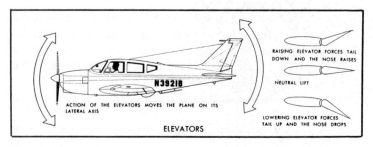

Fig. 1-16. Effect of elevators.

However, with the right thumb up lean the stick to the right. It still points to the side of the aircraft with the up aileron.

So big deal, you say. What's so important about ailerons going up and down? When I'm asked this question I think back to a little scam that Allen Funt presented on *Candid Camera* some years ago. In that sequence, they paid a mechanic to rig the steering on an automobile to do exactly the opposite of what it was supposed to do. If the wheel was turned left, the car went right. Then they schemed to get people to drive it. It was pretty funny. But what if a mechanic did that to your airplane—by accident, of course. The results would probably be disastrous and not funny. If you think that this is far fetched and that it doesn't happen, think again.

In 1977, I was flying freight for an outfit out of Tulsa, Oklahoma. Flying freight even that recently was more like staring death in the eye than an enjoyable occupation. Anyway, our (rather their) company was not famous for the sharpest mechanics in the field. Picking up an airplane from the maintenance facility in Reno, Nevada, one day, I was surprised to find that even with a couple thousand hours of experience I could not taxi the airplane out of a row of closely spaced airplanes. Finally, I did coax it onto a taxiway and proceeded to investigate just what was wrong. When I pressed the right toebrake the airplane would sharply veer to the left and vice versa. The mechanic had hooked up the brakes backwards. The hard part was taxiing back to the maintenance hangar.

If this sort of thing can happen on brakes, it can happen with *any* of the flight controls. It has happened and most often with the ailerons. The lesson here is to check the controls "free and *correct.*" To discover that the controls are not correct in the air leaves one very little time to master the adversity of the situation.

The elevators, so named because they make the airplane go up and down, move the aircraft about the lateral axis (Fig. 1-16). We find the elevator on what most newcomers to flying call the tail. You have undoubtedly noticed that the tail has both a vertical section and a horizontal section. The entire "tail" is called the *empennage*. (It rhymes with fuselage.) The elevator is located with the horizontal stabilizer and is the hinged part of that arrangement. It swings up and down. The elevator and horizontal stabilizer work together to form a single airfoil. The elevator swinging up and down directs lift to the top side or underside of the horizontal stabilizer. When the elevator's trailing edge is lowered, lift is increased on the top side of the horizontal stabilizer. The tendancy for the empennage is to climb. This in turn forces the nose of the aircraft downward and decreases the angle of attack on the wings. The entire aircraft begins a descent.

Whenever lift is decreased on the horizontal stabilizer by lifting the trailing edge of the elevator, the nose of the aircraft is elevated. Hence, the name of the movable surface. Whenever the nose is elevated the angle of attack is also increased. We know that this means that lift will increase and the plane will climb.

Just as the ailerons are connected to the control wheel, so is the elevator. When forward pressure is applied to the control wheel, the trailing edge of the elevator moves down which as we discussed decreases lift on the wings. When back pressure is applied to the control yoke, the trailing edge of the elevator moves upward. This will increase lift on the wings and decrease lift on the horizontal stabilizer.

Take a few minutes here to apply what we have learned about how airfoils to the elevator. The elevator will in fact act just as does the wing. When the camber of the horizontal stabilizer is increased the lift on it is increased.

Another thing one should know is that not all aircraft are designed and equipped with horizontal stabilizer and elevator arrangements. Piper Aircraft Corp. has utilized what they call a *stabilator* for many years. The stabilator is the marriage of the two words stabilizer and elevator. This arrangement differs in that the entire surface is actually an upside-down wing. The assembly is hinged in order that it will move and be able to change its angle of attack and thus direct lift in the appropriate direction. This setup works very well, I have found, with one word of caution. At low landing speeds it is harder to keep the nose high through main wheel touchdown because as speed dwindles, so does the effectiveness of the stabilator. With a little practice, this tendency can be avoided but seems a little more difficult for a student to master in comparison with the conventional arrangement of other aircraft.

Now let's take a look at the rudder (Fig. 1-17). It is probably the least understood control surface by student pilots. This is because everyone is familiar the control wheel in a car and relates that to an aircraft. But the difficulty of relating pedals on the floor that control direction to an automobile is another thing.

The rudder, as we mentioned, is part of the larger arrangement called the empennage. It actually is the movable part of the vertical stabilizer and works just like the elevator, but in a vertical direction. It swings from left to right. We mentioned that the airplane is controlled about its vertical axis by the rudder. This movement about that axis is called *yaw*. It is always coordinated with use of the ailerons because a yawing motion by itself to turn a plane is *very* uncomfortable to passengers. It is something like going around a corner in a car at too fast a speed.

What is so hard for student pilots to understand is that the rudder must be used in concert with the ailerons. What the rudder does is this: It swings the tail of the aircraft out in a turn just at the proper rate to make the turn feel as smooth as if you are sitting in a chair. Without use of the rudder (feet flat on the floor) the tail of the aircraft will try to fall towards the center of the turn. "Fall" here is not meant in the truest sense of the word. Imagine that we are going to make a left turn in an airplane. The left wing dips very slowly as we apply control wheel pressure to the left. But we leave our feet on

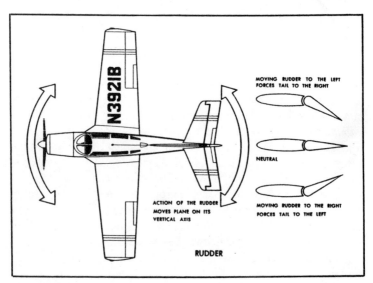

Fig. 1-17. Effect of rudder.

the floor. As we get into the turn, we notice the tendency of our body trying to lean against the left inside wall of the airplane. What is happening is that the tail is trying to move towards the left wingtip or towards the center of the turn. Although we could fly for a while and get used to this odd feeling, most passengers don't appreciate it.

The opposite of not using any rudder in a turn is to use too much. In this case, the rudder tries to move towards the outside of the turn. This is similar to a car skidding around a curve on an icy road. So the answer to using the rudder is to use some, but not too much—easier said than done for the brand new pilot.

Trim Tabs

Trim tabs are control surfaces but not in the strictest sense of the word. They were devised by engineers as labor-saving devices. There are trim tabs on every control surface on every plane. However, they are not all controllable from the cockpit. A trim tab is a small, usually hinged surface on the trailing edge of the control surfaces. They essentially enable the pilot to remove pressure from the primary controls. For example, a pilot is flying along and constantly has to put forward pressure on the yoke to hold the nose down. By using the elevator trim tab, he can turn the elevator trim wheel forward and remove the pressure he has been holding. The airplane will now fly along straight and level.

Trim tabs are sometimes of a mystery—it's difficult to remember what they are doing or which way they are moving. Trim tabs always move in the *opposite* direction from the primary control surface. For instance, in the previous example the pilot was having to hold forward pressure. The elevator was obviously being pushed down. To trim off the pressure, the pilot was holding the elevator down. The trim tab, then, had to be positioned *up* by the trim wheel in the cockpit (Fig. 1-18).

Depending on how expensive an airplane one flies, there will be various combinations of trim wheels in the cockpit. The simplest arrangement (such as in a Cessna 152) is just an elevator trim wheel. Other aircraft have elevator and rudder trim setups and the most expensive have elevator, rudder, and aileron trim wheels.

The aircraft that do not have these trim tabs adjustable from the cockpit have trim tabs just the same. Ordinarily, they are small strips of thin sheet metal and are adjusted by a mechanic. However, any pilot who has been having problems with trimming in any axis can bend these tabs to suit his needs the next time he lands.

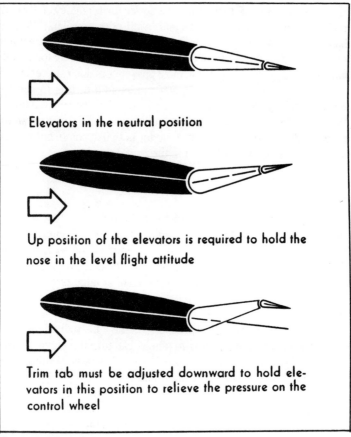

Elevators in the neutral position

Up position of the elevators is required to hold the nose in the level flight attitude

Trim tab must be adjusted downward to hold elevators in this position to relieve the pressure on the control wheel

Fig. 1-18. Effect of trim tabs.

Loads and Load Factors

Airplane strength is measured basically by the total load the wings are capable of supporting without permanent damage. The load imposed upon the wings depends very largely upon the type of flight. The wings must support not only the airplane's own weight, but also the loads that are imposed during manéuvering and turbulence.

In straight and level flight, the wings of an aircraft support a load equal to the plane's weight and its contents. It can be likened to the landing gear on the ground supporting all the weight prior to takeoff. As long as the airplane is in smooth, unaccelerated, straight and level flight the weight or load supported by the wings remains constant. But bank into a turn or apply abrupt or excessive backpressure on the control wheel and the wings take on an additional load. This additional load is brought on by centrifugal force in the turn. This is true whether the turn is in a lateral direction (normal turn) or in a vertical direction such as a loop. There is no way around it. The airplane must be designed to take the additional loads to which it will be subjected. Most passenger-carrying planes, light or heavy, are usually built to withstand loads of three to four and a-half times their normal weight.

The *load factor* is the ratio of the weight the wings are supporting to the actual weight of the airplane and its contents. Another way to look at is to divide the load on the wings by the airplane's weight. For example, at a load factor of two (2) the wings are supporting twice the airplane's weight. At a load factor of three (3) the wings are supporting three times the aircraft's weight and contents.

Each airplane, as we mentioned, has its own design limits. Pilots should be familiar with the stress limits or limit load factor as it is called of the aircraft he flies. Even more importantly, each pilot should be familiar with what conditions will cause inordinately high load factors. The importance of this area cannot be emphasized enough. Load factors are safety items. Now, let's discuss how various regimes of flight affect the load factor.

Effect of Turn on Load Factor. What turns an airplane? Is it the rudder as on a boat? Is it the bank of the airplane? Well, it isn't the rudder, friends and neighbors. It is the *change of direction of lift*. As we have already discussed, lift works perpendicular to the relative wind. It also works perpendicular to the plane of the wing. Now, let's imagine we are going to turn to the left. The first thing we must do is to apply pressure to the left side of the control wheel. This adjusts the ailerons on the wing to cause the left wing to lose lift and the right to gain lift. The plane begins a bank to the left, in other words. Now, think for a moment. If the lift is working perpendicular to the plane of the wing it must also be working perpendicular to gravity. The vector quantity which is working perpendicular to the plane of the wing is pulling the airplane to the left. Vector quantity means that portion of lift that works perpendicular to the plane of the wing in this case. Bear in mind there is also a vector quantity of lift that works against gravity. It is this portion of lift which keeps the airplane flying (Fig. 1-19).

So now we know what makes the airplane turn. However, the relation to this and load factor probably is not very clear unless you are more analytical than average. Any object that travels in a circular path is affected by centrifugal force. A car going around a

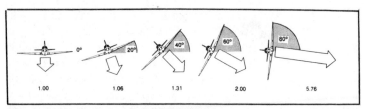

Fig. 1-19. The load supported by the wings increases as the angle of bank increases. The increase is shown by the relative lengths of the white arrows. The figures below the arrows indicate the increase in load factor. For example, the load factor during a 60° bank is 2.0 and the load supported by the wings is twice the weight of the plane in level flight.

curve is acted upon by the same force that acts on an airplane. It is that force that makes one feel like he is being pulled towards the outside of the turn—centrifugal force. So what, you say? Well, the centrifugal force acts against the lift which is acting perpendicular to the plane of the wing. At the same time, the wing must counteract gravity. If the plane is to continue to fly, the angle of attack must be increased. At this point the wing is supporting the weight of the plane and its contents as well as the centrifugal force created by the circular path the plane is taking. In other words, the load factor has increased.

In a coordinated turn (ailerons are coordinated with rudder) pilots and passengers do not feel the pulling sensation as in a car going around a curve. Rather, the first indication of load factor increase is an increase in the feeling of one's body weight. When load on the wings increases, so does the load on a person's body. If we could sit on a bathroom scale during flight, it would register our regular body weight during straight and level flight. If the plane is banked to a 60° angle, the scale would register twice our body weight (Fig. 1-20). I don't think my bathroom scale has numbers that high!

These load factors are also known as *G-units,* a term made famous by the United States space program. A load factor of two is equal to 2 Gs or gravity units. The added weight of turning can be easily sensed through 2 Gs. When the load factor increases to about 3, though, one will notice a sensation of blood draining from your head and a tendency of your cheeks to sag. A considerably greater increase in the load may cause the pilot to "grey out" or even "black out." Losing one's vision is also a possibility, although it will be temporary.

Effect of Turbulence on Load Factor. Turbulence can adversely affect the airframe of an aircraft. The cause of dangerous

26

loads from turbulence comes from strong vertical gusts. The reason that vertical gusts are so dangerous to a plane is that they cause an immediate increase in the angle of attack. This in turn puts a load on the wing which is resisted by the plane's inertia or tendency to stay in one place. There is one effective way to minimize the effects of strong turbulence on an airframe, and that is to slow down to maneuvering speed. Maneuvering speed is that speed at which the airplane will stall before overloading to the wing structure can occur. Usually this speed is in the neighborhood of about 80% of normal cruise speed. We will touch on this again later in more detail. Basically, you should understand that the plane will withstand turbulence of high magnitude at maneuvering speed.

The importance of this one particular subject cannot be overemphasized. For as long as you will fly, you will read, hear, and be instructed not to fly into thunderstorms. The reason is *not* because the lightning will strike your airplane or that the thunder will crush the fuselage. It is because the updrafts and downdrafts will rip your wings off. Also, disorientation can cause a non-instrument rated pilot to get into dangerously steep banks, which you already know

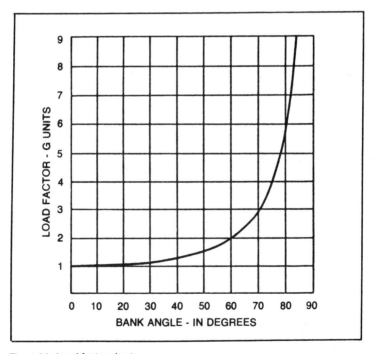

Fig. 1-20. Load factor chart.

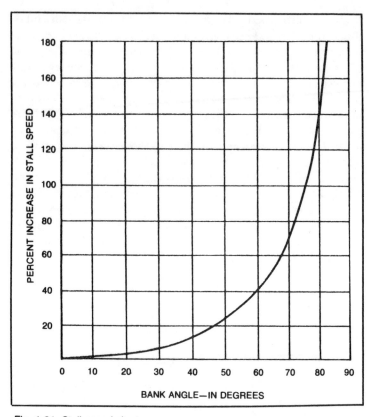

Fig. 1-21. Stall speed chart.

increase the load factor. The combination of the two is usually deadly.

Effect of Speed on Load Factor. This section is basically a rehash of maneuvering speed. Often, though, maneuvering speed is misunderstood by new pilots. The amount of excess load that an airfoil can support is related directly to how fast the aircraft is flying. At slow speeds, the lift available is only slightly more than that needed to overcome gravity and keep the plane aloft. Because of this, load factors cannot become excessive even if controls are moved abruptly or if extreme vertical gusts or shears are encountered. The wing will simply stall first.

At high speeds, the situation is different. The lifting capacity is greatly increased and is in excess of what is needed to keep the plane aloft. For instance, the airplane could be lifted to a somewhat higher altitude by gently raising the nose without increasing power.

There is excess lift. The danger lurks in the fact that the extra lift is available, but the strength in the wings is not there. Safe limits can be exceeded. Because of this relationship between safety and speed, certain maximum speeds are established for each type of aircraft. Whenever the plane flies below these speeds, safety is ensured. The speeds to which I refer are *maneuvering speed* and *gust penetration speed.* These speeds, when adhered to, will allow a pilot to fly the airplane through flight maneuvers which necessitate the use of abrupt movement of the controls. Or, in the case of rough air, a pilot and plane can proceed safely.

You may never see the term "gust penetration speed" in association with the lighter type of airplanes. Usually, maneuvering speed is safe for both abrupt maneuvers and turbulence penetration. I have mentioned it here because in all the basic flying literature that I read when I was learning I never saw it either. There is a slight chance that some of you reading this book will go on to become professional pilots and one needs all the background information he can absorb.

So in summary, at speeds below maneuvering speed, the airplane will stall before the load factor can become excessive. If we fly above the maneuvering speed and abruptly use controls we can exceed safe limits designed into the plane. If you ever think you have possibly overstressed the airplane from a maneuver, have a mechanic check for damage.

Effect of Load Factor on Stall Speed. As we discussed, load factor can be increased by two things in an airplane. Turns can effectively increase the load factor when made at a constant altitude, and turbulence can increase the load factor. We mentioned speed also, but it is more an accessory to the other two causes than a primary cause. The point here is that both turns and turbulence increase load factor and at the same time increase stall speed.

Figure 1-21 illustrates well how bank angle and stall speed go together. Also, check the graph that coincides angle of bank to load factor. The graph shows that in a 60° bank, the stall speed increases by more than 40 percent. In a 75° bank, the load factor is approximately 4 and the stall speed is doubled.

It is extremely interesting to note that the minimum limit load factor for normal category airplanes weighing less than 4000 pounds is 3.8. This value is exceeded in a 75° bank. So the case is clear: Steep banks should be avoided because they are uncomfortable to passengers, they are the possible cause of overstressing an aircraft unnecessarily, and they increase the stall speed.

=Power Off= STALLING SPEEDS MPH—CAS

Gross Weight 1600 lbs.	ANGLE OF BANK			
CONDITION	0°	20°	40°	60°
Flaps UP	55	57	63	78
Flaps 20°	49	51	56	70
Flaps 40°	48	49	54	67

Fig. 1-22. Stall speed vs. flap setting and angle of bank.

Take a look at Fig. 1-22 which compares stall speeds with flap settings and bank angles. You will notice that at zero flap setting the stall speed varies from a low of 55 miles per hour to a high of 78 mph. With the flaps down 40° the speed varies from 48 to 67mph. In each example, the next higher stall speed is an increase of 40%.

Interestingly enough, the increment of miles per hour between each successive increase of 20° bank is not the same amount. For example, on the "flaps up" portion of the chart, the stall speed at 0 degrees is 55 miles per hour. At a 20° bank angle the stall speed increases to 57 miles per hour. This is a difference of 2 miles per hour. Between the 40° bank angle and the 60° bank angle the increment is 15 miles per hour. A glance back at the stall speed graph would show that the increment between 60° and 80° would be about 54 miles per hour.

From the stall speed chart with various flap settings it is readily apparent that the stall speed goes down as the flaps go down. The importance of this may not dawn on you until you actually start flying traffic patterns. The importance is there, however. The stall speed is higher with flaps up and steeper banks. Thus, higher approach speeds are needed in the traffic pattern as well as lower angles of bank. Use 30 degree banks as a maximum and 20 degree banks as the norm. An area that we will deal with at some length later is the turn from base leg to final.

The turn from base leg to final is an especially critical area of flight. The airspeed is generally reduced during this phase of flight.

An overshoot of the runway centerline may cause some pilots to steepen the bank more than needed, bringing the airfoil closer to the stall. A stall close to the ground may be too much to handle with minimal altitude left.

Summary

From the basic understanding of what makes an airplane fly and turn we have laid a firm foundation for using these phenomena in a careful and controlled way. Although anyone can make a fatal mistake in an airplane, those with a good fundamental background of what makes airplanes tick go on to be the safest pilots. It is because knowing what determines the limits of the wing also determines the limits of good operating practices. The safe and prudent pilot operates his or her airplane well within the boundaries of these built-in limits. No normal flight should be an exercise in pushing an airplane to its design limits. Passengers don't appreciate that kind of flight and neither does the airplane. An airplane is not like a horse that you can ride hard and put away wet and the next time you ride it will be mended. Any damage done to an airplane is permanent damage and must be repaired, usually at great cost. So the bottom line is *understand what makes the limits and stay within them.*

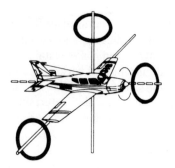

Chapter 2

A Good First Lesson

Waiting in the lobby of the flight school for my instructor to come in from lunch just added to my anxiety. The occasional buzzing by of a Cessna 150 doing touch-and-goes, though, had the effect of whetting my appetite. It made the minutes enjoyable and provided a dreamy quality. But soon my instructor entered and introduced himself. Was I really going to go through with this thing?

In the classroom, he outlined the four forces and what makes the wing fly. I didn't know it at the time, but I didn't miss a thing he said. Most of what we discussed in Chapter 1 is what he relayed to me that very first day. And I have relayed it countless times to others with that burning desire to learn to fly. With an understanding of those two things, a great deal of the anxiety and sense of danger was taken out of the process of learning to fly. After all, I figured, if they know just what makes an airplane fly, then it isn't some kind of fluke that will pick my lesson to stop working.

My instructor kept talking. All the way out to the plane he never stopped. Was he talking because he was nervous? Was he talking continuously to put me at ease or was there just that much to know? My head was swimming with all these ideas. The answer to the questions, I know today. He wasn't nervous. He was trying to put me at ease and fill me in on all of the pertinent facts of flying an airplane. These were facts that someday, the day I soloed, would be called upon. I even would be calling on them daily for as long as I flew. The teaching of the technique of flying might be said not to be the first requirement of good primary instruction. The most important requirement is safety. This means, of course, that one can take off, circle the field, and land. But most importantly, the student must be able to *think* in the air and use his head.

It is wrong to emphasize how fast a student pilot can learn to solo. Almost anyone can be taught to take off and land in just a few hours if he has been started out with takeoffs, turns, and landings. As a matter of fact, they have taught monkeys to do this much. A student with this ability, however, is far from knowing anything about flying. Any student of mine must first prove that he knows a lot more than takeoffs and landings before he or she solos.

The Preflight Inspection

Let's get back to the lesson. Almost no flight begins without a preflight walk-around inspection. The only exception to doing one is when you fly the airplane into an airport and will be flying it back out after only a short time. Even so, at my airline we always check the tires and fuel caps at the minimum.

There is an old joke which holds that, in winter, the thoroughness of the preflight inspection is directly proportional to the temperature. But with a knowledge of what to look for, the preflight needn't be too long. So, what does make a good preflight? It is almost important to end a preflight inspection where it began. The theory behind this is that no item will be overlooked. Another idea is that the student should remain as close to the airplane as possible during the walkaround. The theory behind this is that one must nearly walk over the important items, thus limiting the possibility of missing something. To walk around the airplane from a distance of ten feet or so is useless. About all one can tell from that range is whether or not the airplane has wings. Figure 2-1 shows the normal walkaround procedure for a Cessna type of aircraft. The walkaround for a Piper varies only in that it will start on the opposite side.

Every good preflight inspection should begin in the cockpit. Turn the master switch on and check fuel quantity gauges. Usually three are two fuel gauges, left and right. If the airplane is equipped with electric flaps, lower these at this time. Turn the master switch off. By starting in the cockpit first, you have accomplished two things. You have found out if the battery is active and if additional fuel is required.

Once outside, check the fuel caps after checking the fuel quantity visually. If the fuel gauges indicated a full tank, the fuel should be right near the top of the tank. If the gauge does not agree it would be best to have a mechanic look at the problem. The reason is that most of us tend to trust the gauge because we can see it while

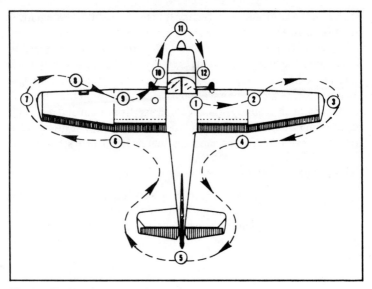

Fig. 2-1. Typical airplane line inspection.

we are flying and we cannot see inside the tank. In our own self-interest, we should have gauges that are reasonably accurate. Also, while peering into the opened fuel cap and into the tank, notice the color of the fuel. Now, to the new student that might sound a little screwy. However, you should know right now that aviation fuel comes in several colors.

The color of aviation fuel is related to the octane rating of that fuel. The way it has been set up is the higher octane ratings have more letters in their color-names. For example, 80/87 octane is the lowest rating. The dye color that has been added to 80/87 octane is red. Only three letters (r-e-d), see? 100 low lead or 100LL as we usually see it on the gas truck is blue and is the next highest rating. Blue has four letters. The other octane rating you might see is 100/115 octane which is leaded. Its color is green (five letters). Don't get frustrated because you don't know which octane the airplane you will be flying requires. Your instructor will tell you.

One other word of caution about fuel: Several airplanes are lost every year because the wrong type of fuel was used. The most prominent mistake is to put jet fuel (kerosene) into the tank of an airplane with piston engines. The color of jet fuel is clear and it smells like kerosene. If the fuel in your tank has no color, it has not faded. It is *jet fuel* and your engine will fail within minutes, possibly after takeoff.

Since you will be standing in front of the wing (with regard to the illustration) at this time while checking fuel, it is also the time to sight down the leading edge of the wing for damage. Any bird strikes or hangar rash will tend to disrupt the design of the wing, and that translates into interrupted air flow. It might not look like much, but it might give us problems.

Walking around to the end of the wing we reach the wingtip. On most training aircraft, the wingtip is a non-structural member of the plane. This means that the wingtip supports no weight in flight. Thus, if you find it cracked (usually they are fiberglass) there is no harm in flying it that way. The important item on the wingtip is the navigation light. It should be intact, although regulations only require it for night flights.

Although I failed to mention it in the cockpit check, the control lock should be removed. This allows easy inspection of all the control surfaces. When inspecting the control surfaces we look for dents or cracks. Cracks can be extremely hazardous, as the control surface may split and a partial loss of control may result. Dents usually do not present much of a problem. Extreme cases of hangar rash, however, must be repaired. Whenever the dents are near the center of the control surface they don't ordinarily cause much disruption in airflow. When a dent is near the edge of the control surface, there is always a possibility that the deformation can result in a jammed control

Besides the overall condition of the control surface itself, the actuating mechanism should be checked (Figs. 2-2 through 2-4).

Fig. 2-2. Checking the hinges, cotter pins, and actuating rod on the aileron.

Fig. 2-3. Hinges and nuts are susceptible to vibration and must be checked each flight. The entire skin must also be looked over.

The actuating rod should be firmly attached and have lock nuts readily checkable. Usually under each aileron there are at least three hinges. The control surface should move about easily on these. Also, at the end of each hinge is a cotter key. Be sure that the cotter key is in place. If it isn't, call a mechanic. It won't take a minute for him to install another.

On the more complex aircraft, geared trim tabs are apparent on control surfaces. A geared trim tab mechanically adjusts itself to the

Fig. 2-4. Checking another hinge and the cable attachments.

control wheel input by a linkage. Check the attachment of this linkage to ensure that it will not come loose and bind the control surface. These are most often found on light aircraft of the type made by Piper.

Back when we did the cockpit check we lowered the flaps. Flaps of the Fowler type are the customary installation on most modern light singles and twins. These flaps normally move backward on tracks as they extend. Although the flaps are probably working properly, check for safety's sake to make sure they're on track. They will also have an actuating rod similar to the aileron's next to them. Make sure the lock nuts are tight and in place (Fig. 2-5).

Next, as we move away from the trailing edge of the wing towards the empennage we should inspect the fuselage. This is where the rule of staying next to the airplane plays an important part in discovering any abnormalities. What we should be looking for is dents or areas of the skin that seem warped or twisted. In all the years I have been flying I have never seen anything like a warp, which would suggest structural damage, but I look very closely anyway. This can be very important if you rent the aircraft you fly.

Another thing to check for along the fuselage is loose rivets. These are another rare item to find but not extremely so. We can identify a loose rivet by a dark worn or greasy-looking circle around the rivet head. Once that has attracted our attention, touch it with a finger and see if it is indeed loose. Sometimes, plain old dirty airplanes look like they have thousands of loose rivets.

As we stroll back towards the tail we might notice a fairing on the base of the vertical stabilizer. Often this is made from fiberglass

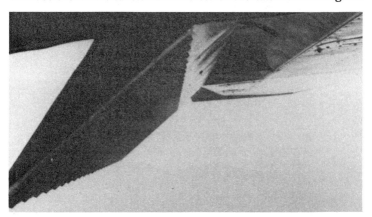

Fig. 2-5. The same things must be checked on the flap as on the aileron.

37

on the more modern aircraft. Usually it is warped from sitting out in the heat and looks loose. Once again, this piece is similar to the fiberglass wingtips. It is not a structural member of the airplane. Its main function is to smooth air around the base of the rudder. Unless it is actually loose, it is acceptable.

Now we should be facing the horizontal stabilizer with our back to the wing and cockpit area. Reach up underneath and feel where the leading edge of the horizontal stabilizer ends. Usually there is an overlap of sheet metal there. Now, right next to the fuselage, grab that overlap with your fingernails and try to pull it down towards the ground. I promise that if an airplane undergoes structural damage in flight, this is the most likely place to discover that damage. It wasn't too long ago I talked to a former student of several years back. He remarked that he never thought he would find any damage, but since his instructor said to always check that area on a preflight, he always had. Sure enough, one day on a rental airplane he was about to fly, he found a loose bond near the fuselage. Now, I'd like to say that personally, I never have found any similar damage, but . . . my instructor showed that to me on my very first lesson and I have always checked it as well.

At this point it is easy to look up and examine the leading edge of the rudder. Check the rotating beacon and watch for antenna attachments that might have vibrated loose. Now, stoop down and pan across the entire belly. There may be several radio antennas down there so check for proper attachment. The one important thing to look for down there is oil on the airplane skin. It probably indicates a leak or excessive oil burn in the engine. To me this would be a definite no-go item regardless what the mechanic or the FBO might say. If it isn't your airplane, rent another one.

On the trailing edge side of the empennage we should check some familiar items. On the movable surfaces check for dents, cracks, and dings. Check the security of actuating rods and hinges. Don't forget those cotter keys.

That brings us to the halfway point of the walkaround. The second half is just like the first until we get to the nose of the airplane. At the nose we must check the pitot-static system. Usually the pitot tube is on the pilot's side of the airplane, unless fitted with two. Check that they are uncovered and clear of obstructions. On Cessna aircraft the static port is near the front of the airplane as well (Fig. 2-6). On Pipers they are under the wing on the back of the pitot tube itself. Your instructor will ensure that you know where the static port is located.

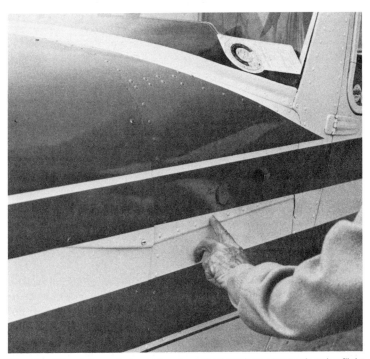

Fig. 2-6. The static port should be clear of all trash to ensure that the flight instruments work.

There are fuel tank sumps under each wing. These are designed to be at the lowest point in the fuel tank. The idea is that this is where water will collect if it is in the tank. Water can get into fuel from two sources: One is condensation and the other is from the pump truck. I have always been taught (and have taught) to drain several ounces of fuel into a collector jar and check it for water in the bottom of the jar. I have read, however, that in some cases that as much as two gallons of fuel were drained before water began to show up in the fuel. Be your own judge as to how reliable this check actually is (Figs. 2-7, 2-8).

Once I am in front of the plane I check the tires. This is especially true of low wing planes. From here we can check tire inflation and condition. If a further check is warranted, crawl up under there and look closely.

Most airplanes are fitted with disc brakes these days. When checking the condition of the brakes, watch for brake fluid dripping down the backside of the tire. Also, note whether the rotor (disc) has been warped by excessive heat from dragging. Brake pads are

Fig. 2-7. Checking the gasoline sumps for color and water contamination.

normally hard to examine. If they make noise while taxiing, they probably need to be replaced.

On Cessna aircraft, some of these items such as tires and brakes are checked as we come to them. On Piper they are checked the same way but the vantage points are often much different. So basically this is a discussion on what to look for on each item.

Most light aircraft have easy access doors to the engine compartment. Oil is usually checked and each aircraft has a minimum no-go oil level. Your instructor will instruct as to how much is proper. Also, inside the engine compartment we should check for the condition of spark plug wires and hoses. Clamps can become loose due to vibration and hoses (such as carburetor heat hoses) will fall off. Also, spark plug wires are occasionally left off after an engine maintenance inspection. Renting an airplane requires one to be more cautious of this sort of thing, as one is never sure when a

Fig. 2-8. Physically check the quantity and color of fuel in the tanks.

plane has just come out of a 100-hour check or an annual. When one owns his own plane these inspections are ordered by the owner and thus he knows when extra caution is needed.

One of the most important things on the front of the airplane, of course, is the propeller. The overall condition of the prop is important and what we first take in. The leading edge is inspected for nicks and dents (Fig. 2-9). The running propeller will "suck" small rocks or gravel up to itself which sometimes can do irrepairable damage. Most dents, however, can be filled out by a mechanic within tolerances. Small dents or nicks often do not bear much thought. Larger nicks are important because, mechanics tell me, a crack can develop and work its way to the other side of the prop in short fashion. This leaves one with an unusable prop and in dire circumstances.

Another item to check while we have our hands on the prop is the engine mounts. Yes, you read that right. We can check the engine mounts with our hands on the prop. Grab one end of the prop and lean outward from the spinner. If the engine mounts are loose, the engine will move inside the cowling. Needless to say, don't fly that airplane.

In cold weather it is always good to pull the prop through a couple of turns. This loosens up the bond that the viscous oil has made from sitting in cold temperatures. The only caution is that the master switch be in the OFF position. We wouldn't want you to tangle with a wild propeller.

Fig. 2-9. When we check the prop we are looking for nicks or dents. By pulling on the prop gently we can also check for broken engine mounts.

The Flight Instruments

Now comes the time we have been waiting for—getting in the airplane and flying. Once inside an airplane, the magic grabs a new student from all directions. Part of the magic is due to the unfamiliarity of all those gauges and switches and instruments. There must be enough stuff in one of those little airplanes to fill the shelves of one complete aisle in a large supermarket. Well, almost. It's tough to think that a pilot must know everything those switches and instruments mean and do just to fly from here to the Poconos. Because of this aura of magnificence, most instructors strive to relieve the complexity of the moment. The best way to do this is to take one instrument at a time. As far as switches go, most instructors don't give them any attention unless they will be used in the first flight. This cuts an awful lot of extraneous information out of the first lesson. Besides, there will be plenty of time later to explore every switch.

We will talk about the various instruments later, as would your instructor at this point. It is the thrust of this chapter to emphasize the flying portion of the lesson. However, it is important to know a few basic things about them. The *airspeed indicator*, usually in the upper left corner of the pilot's flight panel, is read just as the speedometer in your car. It may be knots, but it doesn't really matter; any unit of speed is relative. For example, if your instructor says pull back on the yoke (rotate) for takeoff at 70 knots, that is all you need to know.

As far as the *altimeter* goes, you need to realize that it is read like a clock (Fig. 2-10). If the hands are moving to the right or clockwise, the altitude of the aircraft is increasing. If they are moving we must be flying level.

The *attitude gyro* or *artificial horizon* is used to keep the plane under control when the actual horizon cannot be seen. The main purpose of this instrument during the first few flights is to know how much we are banking to the left or right. There are several ticks around the top of the gauge. Each tick represents 10 degrees of bank. The first tick is equal to 10° of bank. The second tick is equal to 20° of bank and the third is 30° of bank. There are two more ticks after the 30° mark. The first is 60° and the last is 90°. No turns during your first few flights will be beyond 30° of bank. Banks of 30° may seem steep when you are performing them, but they are average turns used by the airlines and anybody else that flies.

Another instrument on the flight panel that will probably be unfamiliar is the *turn coordinator* (or in older planes, the *turn-and-*

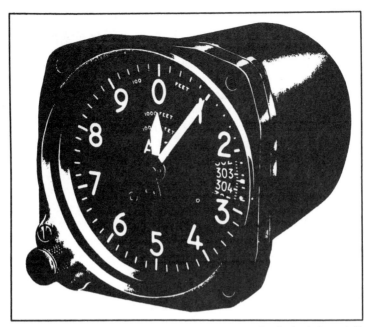

Fig. 2-10. Sensitive altimeter. The instrument is adjusted by the knob (lower left) so the current altimeter setting (30.34 here) appears in the window to the right.

bank). You can tell them apart easily. The turn coordinator has an airplane with wings that dip when you turn. The turn-and bank is just a needle. They both have a ball underneath the main display. This is called an *inclinometer*. This will be or should be an important focus of the lesson—just the ball portion of the instrument.

Usually in the center of the flight panel on the bottom row is an instrument known as the *directional gyro*. It is the same as a compass and is in fact, set to the indication on the compass just prior to takeoff. The thing to remember for the first lesson is that the numbers get larger in a turn to the right. They get smaller as we turn to the left. If you are not familiar with the cardinal headings and their corresponding heading numbers, here they are: North is 0° or 360°. East is 090°. South is 180°. West is 270°. Most turns in the first few lessons will begin and end on these cardinal headings.

The remaining instrument in the main flight instrument grouping is the *vertical speed indicator*. This instrument has one needle and is divided into units of 100 feet up or down. Usually, there will be some large numbers painted on the dial. They are 5, which stands for 500 feet either up or down; 10 which stands for 1000 feet either up or down, and 20 which is 2000 feet up or down.

Once we are settled into our seat and buckled up it is time to start the before-start checklist. Since the type of aircraft determines the items on the checklist it would be futile to detail it here. Suffice it to say that your instructor will get you squared away on where each item is and how to operate it. Finally, at the end of the checklist, you will start the engine. The thrill of that moment can only be equalled by the moment of liftoff to come.

Before taxiing, a brief use of the radio is usually in order. At tower controlled airports a request for taxi clearance is made. It's probably a good idea for the student (you) to start using the radio from the very first lesson. While it is true that there are so many things to comprehend in the first lesson, it is equally true that radio communication is as much a part of flying today as are the wings. I feel the student needs as much exposure to the radio side of things as he can get.

Taxiing

Taxiing to the runway is an experience that you will never forget. Believe it or not, I never knew prior to my first lesson that light aircraft are taxied (guided) by use of the rudder pedals. For a dyed-in-the-wool automobile driver (like most of us), it is a terrible experience to herd that little plane to the runway with bumbling feet. To turn right, place the ball of your foot on the bottom of the right pedal and push. For a left turn, do the same thing with the left. To make a tight, pivoting turn, it is a good practice to hold the toe brake down in the direction of the turn. In other words, if you want to turn to the left, push the toe brake on the left. This is only for *tight* turns. It is extremely bad procedure to use the toe brakes in normal taxiing.

Most taxiways are marked with a yellow line in the center. This line will keep the wings an equally safe distance from obstructions on the side of the taxiway. Your flight instructor will demand that you strive to keep the nose or tail wheel tracking down the yellow line. Wandering from side to side like you had one too many beers for lunch might turn into an exasperating experience. Taxiing an airplane is strictly a case of *less is more*. Make the corrections as small as you think you can. Try to relate it to driving a car down the highway—you don't turn the wheel wildly from side to side. In an airplane, don't use the rudder pedals so much that it looks like you are bicycling your way to the run-up pad.

Another word of caution: Taxi speeds are normally quite slow. Most instructors will tell their students not to taxi faster than a walk. Personally, a speed that slow drives me crazy. It also holds up anybody taxiing behind us. For the first two or three lessons, *do* taxi slow. But after that, pick up the speed to a safe but comfortable velocity. The only place that speed must be monitored very carefully is on a turn. Since most light trainers these days are tricycle geared, it creates a situation where one of the main gears can possibly come off the ground in a too-fast turn. The result may be that the wingtip on one side scrapes the ground—that, in turn, may scrape your wallet.

That sweet and gentle loping of the little engine and the ever-so-faint aroma of avgas weaving themselves into a perfect harmony begin to dupe the brain into an opious state. Taxiing the airplane takes 80% of your concentration. 10% of your attention is tuned to what the instructor is saying and the other 10% just enjoys the whole experience. For me that last 10% has replayed in my mind again and again. Getting to the runway the first time is more fun than anyone would expect.

The Run-Up

The run-up is as much a ritual in lightplane flying as was the wearing of scarves and goggles in the early days of flight. The revving of the engine, the cycling of the prop and the checking of the controls are all a part of the run-up. They all have a purpose. It is the last time that these items will be checked and can be found in good or bad shape before some real demanding decisions have to be made. To skip the run-up in an internal combustion piston engine can be very costly. For example, plugs have a way of fouling at indiscriminate times. If one or two fouled towards the end of flight prior to your next flight, you might not notice it during the reduced power of landing. On takeoff, however, a reduction of power due to fouled plugs could cause a great deal of concern. Even worse, it has caused pilots to push the panic button and do the wrong thing when the situation was not as serious as they thought. A thorough run-up then, can be an aversion to any surprises.

What do we look for in a good run-up? We look for the engine to be in good shape, controls and instruments all in operating condition. Let's be more specific:

When checking the engine, we run the tachometer to a factory predetermined setting. Usually, 1700 or 1800 rpms is the common setting. Once the power is set, we pull the carburetor heat out and if it is working properly—the rpm should drop about 200 rpm. What is happening here should be understood because it can be applied to other areas later. When heated air from the exhaust manifold is routed into the carburetor, it essentially causes the air density to decrease. As a result, the fuel/air mixture becomes excessively rich with fuel. The engine becomes somewhat choked with fuel and the rpm falls off from a normal setting. This same thing happens on hot days without the use of carburetor heat—power is at a loss, and takeoffs become longer.

It also occurs to me that new pilots at this point may not know what carburetor heat is supposed to do. Well, suffice it for now to say that under certain atmospheric conditions (not necessarily cold), ice will form in the throat of a carburetor and constrict air flow. Carburetor heat is applied to melt this ice.

While we have the rpm shaking the airplane and pounding our ears, we next check the magnetos. There are two magnetos which serve as back-ups to each other. On the ignition switch are several positions. One is the normal ON position after engine start and the other two are for running on one magneto or the other. Of course, there is an OFF position.

To check the magnetos, we position the key to one magneto or the other. The rpm will drop off as ignition is not as efficient with one magneto as it is with two. The normal drop is from 75 to 150 rpm. After testing the first mag, reposition the ignition back to the BOTH position. Let the rpm return to the 1700 or 1800 you have set. Then check the other magneto for the same tolerances. Now, there is one other thing to notice while checking the mags and that is the difference between the two mag drops. The maximum difference is usually 125 rpm.

After checking the magnetos, we should turn our attention to the suction gauge. This gauge is connected to a vacuum pump which is the power source for some of the flight instruments. They are all divided differently, so your instructor will clue you in on the proper limits.

After this final check, the engine rpm can be retarded. The next item to be checked will be defined by the checklist. The next item that we discuss here may not follow any checklist but are items that should be noted before taking the runway.

We already talked about controls in the first chapter. When they are checked, take them to the extremes of their travel. Turn it left and full right with the yoke pulled all the way back. Then push the yoke full forward and then check it full left and right. We are checking for the controls to bind in any of the above positions. Full freedom of movement is necessary before we take to the air.

Remember the thumb-points-to-the-up-aileron rule? Well, we should be using it all during the control check. We already discussed how important it is to know that the controls are correct.

The flight instruments are next to demand our attention. Most notably, we must set the attitude indicator for flight. You probably won't be using it very much until later in the program, but the habit of setting it before each flight needs to be started. One thing you *will* be using is the directional gyro. This instrument is set to the magnetic compass which is located above the *glare shield* (another way of saying *dashboard* in aviation lingo). The numbers on the directional gyro should agree with the numbers on the magnetic compass.

The altimeter is probably the most necessary of the three instruments to be set. The altimeter is set in one of two ways. The most common is to set the altimeter to the field elevation. For example, if the field elevation is 110 feet, the altimeter is set to 110 feet. The reason this is that when we return to the airport for landing, we will know how far below us the runway is.

The other way to set the altimeter is to a setting. These settings are derived from ground stations such as control towers, approach controls, and flight service stations. Some fixed base operators also have altimeter settings available and these settings can be received by calling on the appropriate unicom frequency for the airport.

This just about completes most run-ups in trainer type aircraft. The few differences will be due to different airplane types. Those differences will be listed on the checklist and your instructor will keep you informed as well.

The 1965 edition of a book by the FAA called *Flight Training* states that the student should not be permitted to handle any but the most elementary of flight maneuvers on the first flight. They insist that the instructor perform most of the flight and that the student remain mostly as a passenger. I am here to take exception to that thinking. On days of light winds, most of my students were allowed to make the takeoff. Without a crosswind, takeoffs are a fairly simple maneuver, if the student is briefed as to what to expect. The flight *is* mainly for familiarization; however, if it is handled properly you will learn much more than what it is like to ride in an airplane.

So, let's take it from the run-up pad as if you will be performing this first takeoff in your flying career. All takeoffs are made at full power in light trainers. The throttle must be pushed forward positively and fairly quickly. File that away for just a little bit later.

Next, we must be aware that the ground portion of the takeoff is nothing more than a high speed taxi. *"Nothing more than a high speed taxi,"* he says. You are probably thinking at this point that if taxiing is as hard to master as I thought it was, we are liable to veer off into a cornfield or something before we get airborne—and the nearest cornfield is up in Iowa or someplace. Actually, the high speed taxi for takeoff is somewhat *easier* than the taxi to the runway. The rudder and vertical stabilizer align themselves into the wind and on near calm days serve to keep the airplane pointed in the last direction you pointed it. Once you have it straight, it will tend to remain straight.

There is an exception to that last rule of course, as there are to most rules. Your instructor should tell you about the left turning tendency of the airplane at low speeds and high power settings. That's right; an airplane will tend to turn left by itself if the controls are not used against it. This is due to several things; the main one is because the descending blade of the propeller, which is on the right side of the cowling as viewed from the cockpit, produces more thrust than the left ascending side. This tends to pull the airplane to the left. Also, a corkscrew wind from the propeller pushes on the left side of the rudder which moves the rudder to the right and nose to the left. Although these are not all the reasons, suffice it to say that the airplane will turn left without help during low speeds and high power settings. The only way we can counteract this in the cockpit is to use right rudder. The need for right rudder is apparent towards the end of the takeoff roll just prior to liftoff and becomes quite acute at the moment of liftoff. So be prepared to put plenty of right foot down when the airplane lifts off. Believe me, it *is* easier said than done, especially when you have no prior experience and don't have the feel of an airplane. But try it anyway; after a few takeoffs you will have it down pat.

Takeoff

The next thing the instructor will brief you on will be takeoff safety speed. In airplanes, we just don't run down the runway and say, "Well it looks like we are going fast enough to fly." There are design limits and most trainer aircraft don't fly very well before reaching a speed of 70 miles per hour, or knots. Whatever the appropriate speed for the aircraft that you fly that day, you can bet that your instructor will know it.

The moment that we reach that safe flying speed, we fly the airplane off the ground. Some airplanes, if they are trimmed correctly, will actually fly themselves off the ground. Others may require a little tug back on the control wheel. To go faster than takeoff safety speed runs a risk in the other direction. It is possible for the rear of the airplane to start flying and "wheelbarrow" the nose down the runway. This is dangerous at best and so gives extra importance to rotating the nose upward at the proper moment.

Probably the most often-made mistake by a green pilot is to rotate the airplane too strongly. The idea should be to raise the nose of the aircraft slowly. Usually, a good way to know how to do this properly is to look outside at the cowling. Bring the cowling of the engine up to the horizon. This will be close to normal climb attitude for any trainer. Remember to raise the nose gently. You will be surprised at how light a response the controls call for. If you are afraid of the wheelbarrowing tendency, then fear not. If you will add a little back pressure at the moment of rotation, the nosewheel will lift off the ground smoothly and put all of the weight of the aircraft on the main gear and wings. Soon, the wings will whisper away to the clouds.

Now, the briefing is over and we must prepare for taking the runway. The thing to do now depends upon the type of airport we are flying out of as well as instructor preference. If you are taking your first lesson at a tower controlled airfield, you will need a takeoff clearance from the tower. At an uncontrolled field, where there are no tower controllers, we need to make a clearing turn. This is done by increasing power and pivoting with one wheel brake pedal down in a circle. The direction that we turn depends upon where the traffic pattern for landing aircraft happens to be. Of course, your instructor will know what is standard for this runway and airport. The object of the turn is to visually search downwind, base, and final approach to the runway for any arriving traffic. If the pattern is clear or takeoff clearance is given by the tower, we may take the runway.

Now, as we take off, all the things that the instructor just said will be running through your head like wild banshees. Don't worry, he doesn't expect you to remember everything and do the takeoff by yourself. Most instructors will be closely guarding the controls and talking you through every aspect of the takeoff.

Taking the runway, we must line up with the centerline to provide the maximum distance in which to maneuver. If you are flying out of one of those romantic grass strips like the good ol' boys used to do, then try to find the middle by noticing where the greatest amount of wear has occurred.

The moment has arrived, and how you will savor it for a lifetime! Even if you never have the opportunity to fly again, this first takeoff will live on to the end of your days. The instructor is talking to you now, leaning over and getting right in your ear. Feed the power in quickly and smoothly. Don't jam it now. We are already rolling. Before you know it the airplane has reached 40 miles per hour and is accelerating fast. It is like racing a car through a quarter mile course, but at the end we won't slow down, we will fly away. By now, your first correction on the rudder pedals is needed. Whoops! Easy now, they are a great deal more sensitive than when we were taxiing. Needs a little right rudder, doesn't it?

The airspeed indicator reads 70 miles per hour or knots. Once again your instructor is right in your ear. Ease back on the controls and, by golly, you are flying. It sure needs that right rudder. If you are able to take your concentration off flying for a second, you could see that the plane wants to wander left. Unfortunately, if you are like most of us in an unfamiliar situation like flying you will be busy holding the airplane up. That's right—somehow in our minds we will feel that the success and furtherance of our life depends on *holding that airplane up.* Really folks, it will fly if we take our hands off. On my first flight I grabbed the yoke so hard that my left hand went to sleep right after takeoff. The instructor noticed my death grip and told me let to let loose. The airplane flew, and kept right on going as if I weren't even in there. Shucks, I thought this thing needed a pilot. That was my first experience with how easy and safe flying actually is.

Climbout

After the takeoff, the pilot climbs his airplane to a height of 400 feet. At this altitude the first turn is begun. In most patterns (traffic) a standard left turn is required for takeoff and departure from the pattern. Once again, the instructor will know which departure is necessary; left, right, or straight ahead. At a controlled field, the controllers may instruct you to make a certain turn. Nevertheless, it will be your first turn to execute in an airplane.

After you have skittered through the turn, then you will be asked to make another one. This one will be a 45° change in direction to leave the pattern altogether. Besides the mechanics of the turn, most students have a problem knowing how far to turn. In other words, they are not sure how far or how long it takes to turn a plane 90°. One of the easiest ways to accomplish this for a beginner is to look out the wing in the direction of the turn. This will be 90°

from the nose (or close enough). The student must pick a prominent landmark in the direction desired and then turn the airplane to face it. When the object appears out the windscreen—which is another word for windshield—the turn is complete. Sounds easy, doesn't it?

After this "pattern leaving" turn is completed there may be one more to execute which will take you out to the practice area. Once you have the airplane headed for the practice area, you will have a chance to relax for the first time since you taxied out. There are things to do, but they won't take as much concentration. The airplane should be climbing at its optimum speed. As we mentioned, that airspeed is usually the same as needed to set the top of the cowling on the horizon. If your nose is set on the horizon, the climb airspeed will stay constant.

Trimming

The trouble with keeping the nose here on the horizon is that it begins to tire your arm from holding back on the controls. This is your first experience with the trim tab. It's important to keep the control wheel in your hand during the entire trimming process. Without doing this we cannot feel if the pressure has been removed or even reversed. Let's discuss the process.

To most of us, starting out the trim tab is a mystery. We have already discussed how it works aerodynamically. How it works mechanically is another thing. Don't be disappointed if at first you use it wrong. About half of all students do it wrong the first time. Of course, if you pay attention here you won't be in that lower half. The best way is to remember that the trim wheel is attached to the nose of the plane. If you want to hold the nose in a climb position, pull the nose up with the wheel. In this case, roll the trim wheel backwards away from the nose. If you want to hold the nose in a low position below the horizon for a descent, roll the trim wheel forward towards the nose. Trimming the aircraft for level flight can be a mix of the two. If you are leveling from a climb attitude, you will probably have to roll the trim tab forward and the opposite is true for a level-off from a descent.

Of course, what has been said so far tells us little about how to trim the plane in actuality. It is a matter of feel. I always told my students that they must trim the pressure off the control wheel. First, hold the yoke in your hand and lower or raise the nose to the position necessary. If the airplane is out of trim, you will have to hold pressure against the wheel in one direction or another. If holding back pressure (holding the nose up), the trim wheel must be moved away from the nose to "hold the nose up." The converse is true as well. Now, begin with large adjustments and taper off to very small fine adjustments. When the airplane remains with the nose attitude that you have selected, the trim job is finished. You can expect the nose attitude to vary a little. It's just the way the aircraft will respond to changing air currents and requires us to "ride herd" on the controls. It gives us something to do on long trips.

After trimming the airplane for climb, we have one other task to take care of as we fly towards the practice area—keeping the wings level. This is not such an easy thing for the beginner to do. Bascially, the airplane is stable enough to remain with the wings level, but if your first flight is during the summer or after a winter cold front, the air is likely to be slightly bumpy. These bumps alone will cause a pilot to stay busy keeping the wings level.

A new student has trouble keeping the wings level because the "outside references" are not clear to him. As an experienced pilot I use only the references straight in front of me, such as the cowling or windscreen, to maintain wings level. These actually are the best and you will learn to use them in time. For the first lesson, however, it is generally easier for the new pilot to use the wingtips for a level reference. In high-wing aircraft the horizon is below the wings. In low-wing airplanes the horizon is above the wings. In either case, we must compare the left wingtip to the right wingtip. If the horizon is an equal distance above or below the wing—depending which type of airplane is being flown—on both sides of the aircraft, the wings are in fact level.

Outside the plane to the front is also the horizon. The cowling is ordinarily a handy reference. In most single-engine aircraft the horizon will appear to be two to four inches above the cowling. The thing that throws most of us when we are starting out is the curvature of the cowling. Most students have trouble understanding that the horizon can be lined up with the cowling if it is used as a tangent line to the curve. Several students of mine in the past have tried to lay the horizon right on the cowling at its least curved (read that flattest) part. This results in a nose up and left-turning condition. Once again, the main idea is to use the horizon as a tangent so that the center of the cowling is about two or four inches below it, as we mentioned.

So now we have the nose trimmed up and the wings looking level as a Texas sunrise. The remainder of the trip to the practice area will be in the climb. As we arrive in the practice area we will begin the level flight procedure. First, we bring the nose down to a level position (two to four inches below the horizon). Then we trim the forward pressure that we have been holding off—easy does it.

Believe it or not, what you are about to do is the most important flight maneuver you will ever learn. Straight and level flight is indeed a flight maneuver. It may not hold the excitement of a loop or roll or stall, but it *is* a flight manuever. Without learning to properly execute this maneuver, all others will probably fail. Once we are level we must begin to scan several indicators as to how well we are doing. Those indicators are the wingtips, cowling, and altimeter. The object here is to continuously scan between those three. It will probably be easiest to fixate on the altimeter, but this is the worst thing you could do. At this point you are training to be a VFR-rated pilot. That means you will rely on visual clues outside of the cockpit. Although the altimeter must be included in the scan to see how we are doing, it is wrong to rely on it too heavily. If you become thoroughly familiar with the outside references, it becomes possible for you to maintain altitude quite well using them alone. The altimeter should be used as a cross-check and nothing more.

After the nose has been lowered, we must reduce the power on the engine. We should leave the power at the climb setting for a few seconds and let the airspeed increase in order to make final trimming easier. One of the easiest memory aids to use for the trimming of an aircraft to straight and level is "pitch, power, and trim." This is probably backwards to what I just described a few sentences ago. The truth is somewhere in between. In actuality, the nose is usually lowered simultaneously with the reduction of power.

The easiest way to stop a climb in any aircraft is to reduce the power. Airplanes are able to climb only because of the excess thrust that an engine can supply. Thus, to stop an airplane from climbing the throttle should be retarded.

As the power and the angle of the nose are reduced it becomes time to begin the trimming exercise. Remember, as we described it before, the pressure on the control wheel is held by hand and the pressure is rolled away by the trim wheel. During all of this, the pilot must direct his attention outside to the visual references to properly set the trim.

On most trim wheels you will notice a scale or center mark. Some students will try to memorize the position of the indicator. The thinking here is that every time that straight and level flight is required, the indicator can be placed in the same position. *This won't work at all*. As altitudes and the weight of the aircraft change from flight to flight (or even during the same flight) so does the trim requirement.

Now that we have the airplane established in straight and level flight, the instructor will give you a minute or two to relax and let your mind catch up with what has taken place so far. For a minute or so you can concentrate on the outside visual references and watch out the window at this new and spectacular view on life.

Confidence Demonstration by the Instructor

The confidence demonstration may not be standard with all instructors. It always was with me, however, regardless of the student's background. Everyone needs to have some confidence in the airplane he is flying to alleviate a nagging, sometimes unknown fear. For example, a student may be the son or daughter of an airline pilot. Under such circumstances the instructor may believe that they may have handled the controls of an aircraft many times. This sort of thinking may be completely out of line and short the student in an area that is of foremost importance and that is *confidence.* If the truth be known, many sons or daughters of airline pilots may be under some duress to go and learn to fly. In these cases, the student's apprehension must be eased or he/she must be eventually culled out.

The easiest method for easing apprehension is to sit through a confidence demonstration. After dealing with the first takeoff and the climb over to the practice area, most students need a break away from the stress of trying to fly an airplane for the first time. Believe me, when my first instructor took the controls for a few minutes after that first takeoff, the relief was welcome. It will be for you as well.

With most students there will also be some apprehension when the instructor takes the controls. When my instructor did this I thought, "Boy I sure hope he isn't going to show me how well he can fly and turn this thing upside down!" Well, he didn't. As an instructor, I never did and your instructor won't either. Why? Because he wants you to enjoy the whole flight and return for another lesson.

Now, just what is a confidence demonstration? It is a series of facts delivered by the instructor in such a way as to point out how the airplane reacts to certain stimuli. It also is to underline the great engineering and reliability inherent in planes of today.

Did you know that airplanes climb due to excess power? For instance, X amount of power keeps the airplane flying straight and level. If the power is increased the nose will rise and the plane will climb. This is the way I begin a confidence session with a new student. At this point, the throttle is pushed in and the student's attention is called to the nose of the airplane. As the sound of the engine becomes louder, the nose begins to lift. The next thing on every first-timer's mind is what happens when the engine stops. Usually, the instructor will preface the forthcoming event so that the student is not taken by surprise. The throttle is pulled to idle and the instructor will explain that although there is no power now available, the airplane continues to fly. Granted, it is beginning to lose altitude gradually, but the innate fear of winding up on the ground instantaneously has been averted.

Next, it is important to understand how stable the airplane is designed to be. I always began a shallow turn at this point. Telling the student ahead of time as I release the controls. The airplane in its turtle-slow turn begins to slowly right itself. Once again the plane will be flying straight and level.

Now the airplane may be placed in a medium banked turn. A medium bank is one that is from 25° to about 38°. The emphasis here is again on stability. An airplane in a medium-banked turn will tend to stay in that medium bank. It will neither increase nor decrease its angle of bank. The importance of this demonstration is that the airplane, through design, helps a pilot to fly properly and easily.

The next maneuver on the confidence session does tend to make a few new students uneasy. The most likely reason for this is that it is hard for a new flier to imagine how he would get into a situation such as this one I am about to describe. Disorientation is an enemy of all who fly. It doesn't matter that an airline pilot has 10,000 hours or more, he can become susceptible to vertigo or become disoriented under some conditions. It is, however, a great deal more likely that a new pilot would become disoriented. Whenever a pilot becomes disoriented, there is a greater danger that control of the airplane will be lost. In such a condition, a rapid meeting with the ground may be imminent. For the new pilot with little experience, there may be one intelligent thing he can try. Turn loose of the controls altogether.

To demonstrate the plausibility of such an event, the instructor may put the airplane in a steep descending spiral. Turning the controls loose after the airplane is established in this strange maneuver, you won't believe your eyes. The airplane will begin to slowly right itself! At some point it will reach a familiar attitude and the pilot will be once again master of his own fate.

I agree that this may be somewhat uncomfortable for the first-time flier. If the student goes home and thinks about it, he will realize how much safer flying is than he had ever imagined. A good first lesson should not go without a confidence-building session.

Now, It's Your Turn

As with all learning processes, it becomes necessary for a student to take the bull by the horns and try a project. Beginning with a trimmed airplane the student can concentrate for a few seconds on the fine points of straight-and-level flight. The next thing to master is the level turn. You won't actually *master* it this first lesson, but you will learn the mechanics of performing the maneuver. The level turn has two very important aspects to be mastered. These two areas are back pressure on the controls and the centering of the ball in the inclinometer, which is below the turn-and-bank.

As the angle of bank increases in a turn, the amount of lift directed vertically away from the ground decreases. This means that in order to maintain altitude, the angle of attack must be increased. This is done by adding back pressure to the control wheel, i.e., pull back on the wheel. Ordinarily, banks of less than 30° take little or no back pressure, and that is the type of turns you will be making this first time out. So, little back pressure will be required. Just the same, *some* will be needed.

The ball is the other consideration for properly executing a turn. It is also the most intangible item on the panel and thus the most complex for most students to assimilate into their learning and use. I have flown with many pilots who have flown for years and *still* don't have it mastered. Believe me, it is one of the marks of an excellent pilot.

To review, the ball is the instrument which lets us know if the proper amount of rudder has been applied to make the turn a coordinated maneuver. You see, the tail of an aircraft in a turn can swing around to follow the wings either too slowly (not enough rudder), too quickly (too much rudder) or just right (and the ball will be centered). Therefore, the idea is to keep the ball centered during a turn (Fig. 2-11). An interesting point is that more rudder is applied during the entry into a turn than after it has been established. The best way overall to know if the turn is properly coordinated is to glance at the ball. The question that so many new pilots ask is "How am I supposed to watch outside and perform the maneuver and look inside at the ball, too?" The idea is to glance back and forth. This is not easy for the new pilot, but it is necessary, though, because it is the way most flying is done. That is, look out and then check things within. Of course, there are other ways to know if the airplane is coordinated with the proper amount of rudder. We will discuss only one of these now, because on the first lesson a student wouldn't be expected to pick up on the others.

If a turn in an airplane is not coordinated, there is a certain feel to it. The passengers know that it isn't right and the pilot knows it is not correct. That feeling is called *kinesthesia* or the body's sensitivity to motion. If the turn is not properly coordinated, your body will feel like it is leaning one direction or another. If the turn is properly executed, the feeling will be the same as sitting in a chair on the ground. All of your weight will be directed towards the floor of the aircraft. One of the ways I learned to know when I was uncoordinated in a turn was that I was leaning on my instructor or he was leaning on me. Since he didn't like that much, he would glare at me and the lesson was soon learned: *center the ball.*

As for back pressure during the turn, forget it for this lesson. The idea during this lesson is that the mechanics of flying are realized. You should be aware, for example, that rudder is needed in a turn and that power makes the airplane climb. To expect yourself to master any of these control inputs this early in your flying career is to have unrealistic goals.

After attempting a few of these turns, the instructor will get you headed back towards the aerodrome. You may be surprised at how fast time has passed. To do all of this, a good half hour has probably elapsed.

Your instructor may ask you which way to go back to the airport. This is to see how well you might stay oriented. If it is because he doesn't know, get another instructor. Seriously, it is sometimes an indicator that student pilots with high aptitudes for flying have remained aware of their surroundings throughout the lesson. If you can find your way back to the airport, give yourself a

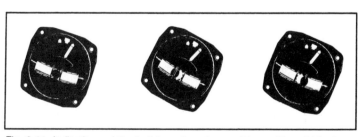

Fig. 2-11. Indications of the ball in various types of turns. The middle figure indicates that too much right rudder is being applied. How about the picture on the right? Which is a slipping turn and which is a skidding turn?

pat on the back. If you don't know which way it is, don't worry over it. I didn't know which way it was either.

The Traffic Pattern

Upon nearing the airport, the instructor will maneuver the plane or have you maneuver the plane for entrance into the traffic pattern. At this point it is not necessary that the first-time pilot comprehend all that is happening in preparation for landing. It is a good idea, however, to at least know what a traffic pattern is and how it works. A traffic pattern is used for landings or takeoffs. It has been devised to keep one airplane from running into another. Patterns work well at uncontrolled airports (those without a control tower) and at controlled fields. The traffic pattern is a rectangle with the runway forming one of the sides. The side of the pattern that contains the runway is considered the *upwind* side and the opposite side is known as the *downwind* side. The downwind side is where most traffic patterns start.

If you are following this so far, you might have thought far enough ahead to see that there can be two traffic patterns to every runway. One would be on the left side and the other on the right side. Left traffic is considered normal. We will be flying left traffic if the runway is to the pilot's left. All ensuing turns to the runway will be left turns. The same is true for right traffic. The runway will be off to the pilot's right. All turns in the traffic pattern will be to the right.

There is a great deal more to know about flying traffic patterns, such as how to depart or enter them. At the time of one's first lesson it is not necessary to know all these facts. The instructor is there to do most of the work and thinking. It is best to rely on him and follow him through on any techniques he may be attempting to teach you this very first lesson.

Landing

After entering the traffic pattern, the instructor may be busy talking on the radio reporting the position of the aircraft in the traffic pattern. He may even ask you to handle the radio, in which case he will explain what to say. This is not the important thing, though. Things like pulling the carburetor heat abeam the runway threshold, and adding flaps as the plane slows down, are the important items. The student is not ordinarily expected to remember or even participate too much in this process the first time around. What you should glean from all this is that things are done at specific points in the traffic pattern or at specific speeds on the airspeed indicator. This is the way proper approaches and landings are made. Usually, if a normal sequence is developed, the landings are all normal. When we start varying the plan from the normal sequence of things, we introduce room for error. This point can be very important in emergency circumstances. What you begin learning here should be carried throughout your entire flying career, even in the most dire circumstances.

Now, let's say we are on the final approach to the runway. This means, by the way, that the runway is directly in front of the airplane and that the airplane will have to turn no more before landing (Fig. 2-12). The landing is a process by which the airplane is made to fly slower and slower until the ground is placed firmly beneath the wheels. On final approach, the airplane is held to one speed as close as turbulence will allow. Then, once over the runway, the airspeed is bled off until a landing is achieved. Depending on the type of landing your instructor demonstrates, you will see one of several things. If it is power-off from the very start, the airplane will glide to the runway. Over the runway, the nose will be raised imperceptibly slowly and the plane will touch down on the main wheels first.

If the instructor demonstrates a full flaps down, power-on approach, the process will be slightly different. A slower airspeed will be held on the final approach. Over the runway, the nose will be raised quite a bit from the nose-low position of final approach. Incidentally, the nose-low attitude is characteristic of a full flap approach. Once the nose is raised, or while raising the nose, the power is slowly and smoothly retarded. The airspeed disappears and the plane will touch the ground, mains first.

Following touchdown, the nosewheel is gently lowered to the ground and brakes are smoothly applied to make the turnoff to the taxiway. Once the nosewheel is on the ground, the airplane is back to taxiing—remember, driving with your feet. After the plane has slowed enough, the instructor will turn it back over to you and let you practice taxiing once again. You may be surprised at how much easier it is already. This is because of the training you received in making turns in the air with the addition of rudder. Things are beginning to gel already. Swing the plane into its appointed parking spot and the best time of your life is through.

The propeller is still spinning and the airplane seems as if it wants to keep flying as much as you do. But we must shut 'er down and wait for another day. Shutting down an airplane is not nearly as easy as turning the key off as we do in a car. First, we must go to the

Fig. 2-12. The view on final approach.

shutdown checklist and accomplish all things listed there. It is different for every type of plane. Basically, though, the engine is killed by pulling the mixture control all the way back. This starves the engine of fuel and empties the carburetor. After the prop has stopped spinning, the key can be turned off. The reason that we do things in this order is to keep from fouling the spark plugs. Aircraft spark plugs are not fouled any easier than automobile spark plugs. It is much harder, however, to pull over to the side of the road and check the problem out in an airplane. Thus we make sure to burn all the fuel that we can out to ensure longer spark plug life and safer future flights.

After the key is off, we must check that all electrical switches are turned off and that, most importantly, the master switch is in the OFF position. Believe me, there is nothing more disappointing than wanting to fly so bad you can taste it and finding the airplane you have reserved has a dead battery. So make sure that master switch is off!

Climbing out of the cockpit and walking inside you may find that a great deal of strength in your legs has disappeared. This is the first indication you will have of how engrossed you were with the flying. Nervous energy has sapped your body of its strength. There is an equalizer, though. When the instructor hands you your first logbook with the first hour of flying experience you will walk on a cloud the rest of the day.

Chapter 3

Learning the Basic Flight Maneuvers

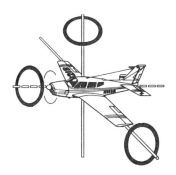

A student came to me by way of another instructor. The instructor, employed by the same FBO I was, had been having trouble getting the student ready for his check ride. It seems that his basic flying maneuvers always fell just short of the recommendation he needed to take a check ride. This situation was not extraordinary; we often traded students who were having trouble. Often another viewpoint introduced by a fresh instructor would be all the student needed to start his learning curve upward again. The interesting thing about this student was that he could explain the maneuvers perfectly. His flying, however, was just short of being good enough.

The student and I went out for a flight. As predicted, his turns were ragged in entry and rollout. His stalls were uncoordinated and bordered on spin entries. After riding through the whole routine with him, I was perplexed. Accidentally, I told him to just fly straight and level while I pondered the problem. As he flew on I noticed that he was looking out of the left window more than anywhere else. Now, it isn't that unusual for a pilot to look out the left window. Closer inspection, though, revealed that he could barely see out the forward window. This student was just a hair too short to see clearly over the glareshield. The result was that the student was having to shift his references to where he *could* see— the left window. So, how does this effect his flying? The answer is in his example of straight-and-level flight. By looking out the left window the pilot unconsciously applied pressure to the controls to the left also. His straight and level flight was not quite level. In fact, the left wing was down or dragging most of the time. This little characteristic carried through all of the student's flying.

Returning to the airport, we looked up his instructor. I told him what I thought. The three of us decided that the student would go back to his original instructor and this time carry a small sofa pillow on subsequent flights. The student's flying improved remarkably and soon he was a licensed private pilot.

The story above tends to prove at least in my mind that straight and level flight is an important foundation of flying. Without a good basic foundation it is nearly impossible to build anything of permanence. This is particularly true of the pilot and his abilities. If this one area is neglected, it becomes easy for a student (or even a licensed pilot) to become disenchanted with his own flying performances.

Visual References

The best method for learning straight and level flight has already been mentioned in the previous chapter. For the beginner, looking at the wingtips and noting their relation to the horizon is easiest. Unfortunately, it is not the best *permanent* technique. This is because as maneuvers become more complex, they require that more attention be given to the instrument flight panel, and the wingtips must be neglected. Therefore, after the first lesson it becomes important that one begin to pick his own reference points. My favorite reference is the cowling, because it is right out front. As we mentioned before, the cowling on all planes is curved to some degree. This can give the beginner some problem in matching the horizon, which is flat, to the cowling. From self analysis I have found that first of all the horizon is always some fixed distance above the cowling. To find this distance, we must level the plane so that we have a zero rate of climb or descent registered on the vertical speed indicator. When we have achieved this, it becomes evident that we are maintaining a level altitude and a level attitude with respect to the lateral axis. At this time, we must look out at the cowling and horizon and memorize how it looks. For instance, the horizon should be about three inches above the cowling. This distance varies a little between each pilot due to the pilot's height. Once one has established his own personal distance mark, it will remain the same flight after flight.

The one thing that the beginner should remember at this point is that what we have just solved is the method for keeping the nose in the correct *pitch* attitude. We have not done anything about learning to keep the wings level. The wings are kept level by checking the wingtips as we mentioned in the last chapter. Soon this becomes a problem, however, with the addition of maneuvers that require the pilot to maintain pitch as well. The best way to maintain pitch and level wings is to combine one reference. Hence, we use the cowling outside again or rather we judge both attitudes from one reference. Since the cowling is curved, it presents the same problem for the beginner. I have had some new students trying to lay the horizon parallel to the cowling's slant away from center. This results in a turn, unfortunately, usually to the right. It becomes necessary for the pilot to use his imagination. The horizon must be placed above the cowling to represent a tangent line. A tangent is a line that touches a curve at only one point. It is very much like a yardstick balancing on a basketball. The basketball in this case would represent the cowling and the yardstick represents the horizon. The only difference between the two is that the horizon in the airplane will be about three inches *above* the cowling.

Now that we know where both references are for straight and level we combine them to make the job easier. For example, if the horizon is higher than our usual two or three inches above the cowling, the nose of the airplane must be too low. The converse is true. Also, if the tangent line in front of us, which is scribed by the horizon, is not level or straight across the windscreen, then one wing must be lower than the other. Which wing is low is ordinarily obvious. The tangent line on that side will be much above the ordinary position. More ground will show in the windscreen on the wing low side. It may sound complex, but really is not. On close inspection and with retrospect after a flight you will be able to understand what all this was about. It is strictly a lot of words to describe a simple procedure.

Medium Banked Turns

After practicing straight and level for an adequate period of time, your instructor will start the process of teaching you medium banked turns. You may not have done any of these on the very first flight because they require a little more finesse than shallow turns. Medium banked turns are those turns having a bank of less than 35° to 38°. They are in fact, the most-used type of turn. This is because they can change the direction of the airplane fairly quickly without making the passengers uncomfortable. Most airplanes commonly use 30° banked turns.

The finesse I mentioned is necessary in the application of back pressure. Back pressure is the term that we use to describe the pulling back on the control wheel. Pushing on the control wheel is known as forward pressure. At any rate, it takes a little bit of time to develop a feel for the back pressure required for a turn.

A medium-banked turn is begun the same way any turn is begun. We apply pressure to the left or right on the control wheel similar to that we use in driving a car. The wing will begin to lower almost immediately as we maintain the pressure. At this point we must monitor the "tick" marks at the top of the attitude indicator. There will be three marks to either side of the center mark for medium turns. The first represents 10° and each successive mark is an increase of 10°. Let's say we are going to execute a 30° banked turn. We will apply pressure until the wings have rolled to a 30° position on the attitude gyro (Fig. 3-1). At this point it becomes necessary to add a little pressure in the opposite direction to stop the roll into the turn. Also we must use a very slight amount of pressure against the turn to overcome any tendency the plane might have to roll into a steeper bank.

As we reach the 30° mark, a marked reduction in lift is noticeable. The plane wants to drop its nose to pick up speed and bring the lift back up to where it was trimmed in straight and level flight. If we are to maintain our altitude throughout this turn, however, we cannot allow the nose to pitch down. We will lose altitude. The correction for this tendency is to increase back pressure on the controls. The amount required is unknown to a new student. Therefore, the student will hunt around for the proper amount. There is an easy way to sidestep this problem. It is up to your instructor's technique, however. Before asking you to try one of these medium-banked turns, he should demonstrate one or two to you in the same direction that he will ask you to try one. While he demonstrates one, the new pilot should concentrate on the visual references outside the cockpit—the very same ones that were used for straight and level. Now you should be beginning to see why straight and level references are so important. When we roll into a turn, the horizon dips to one side or the other. If we don't apply the back pressure at the 30° point the nose will lower and that should be apparent because the distance between the horizon and the cowling will have increased with the lowering of the wing and the nose together.

What is important to note at this point is the attitude presented to you for a level turn. In a Cessna 152, for example, the corners of the cowling will come up to the horizon (Figs. 3-2, 3-3). In every turn of 30° the reference will be the same for the same type of aircraft. Thus, we should memorize that reference. When you try

Fig. 3-1. The instrument indications for a level turn 30° bank to the right.

your first turn, try to break it into three main and easy steps. *One*, roll the airplane to the 30° mark. *Two*, add pressure against the turn to stop the addition of any more bank. *Three*, add back pressure to bring the nose up to the height demonstrated by the instructor. Using this 1-2-3 method, a good first attempt can be recorded.

As if just banking the airplane and holding the nose in the proper pitch position weren't plenty to think about, there comes

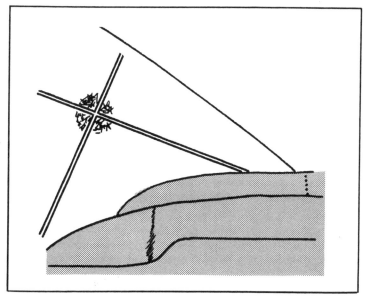

Fig. 3-2. Outside visual references in a 30° bank to the left.

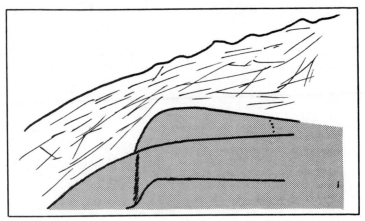

Fig. 3-3. The outside visual references for a right-hand 30° bank.

more. All turns must be coordinated. That means we must use rudder. Remember, rudder is the control we need to make sure the tail follows the wings around in a turn at just the right rate. Rudder is always applied at the very beginning of a turn. When pressure to the right or left is applied, corresponding left or right rudder must be applied. The rudder pressure applied at the entry into the turn is always more than that which is needed after the turn is established and under way. In some aircraft, no rudder at all is required after the bank is established in a medium turn. This is as much a function of design as it may be a function of how well the airplane is rigged.

The amount of rudder required is also an unknown quantity to the fledgling pilot. Fortunately, there is the inclinometer on the flight panel for reference. The ball is the gauge for the amount of rudder required. For nearly all normal flight situations, the ball should be centered. If the ball is to the right of its center marks, then we need to push right rudder. The rule here is *step on the ball*. It works as if the ball were under your feet. A step to the right moves the ball to the left and vice versa.

The utmost basics of flying are simple. There are only four fundamentals for the student to master: straight and level, turns, climbs, and descents. Climb, descent, and turns are the most frequently used basics in maneuvering. If a student is able to do these well after developing a "feel," instead of just using or relying on mechanical movements of the controls, any more complex maneuver can be learned quickly. All that is lacking are the student's visual cues and conception of the new maneuver before he can accomplish it.

Climbs

Although climbs were the biggest part of the first flight, most students lack an understanding of the correct way to perform climbs. Climbs require the proper outside references just as do turns and straight and level. In climbing straight ahead, the reference is usually something like bringing the nose up to the horizon. When we say nose we actually mean the top of the cowling (Fig. 3-4). The cowling is always ready for a handy reference and we will always make good use of it as we learn to fly. In a Cessna 152, for example, placing the nose on the horizon will just about give the best rate of climb airspeed. The best rate of climb speed is the speed at which the airplane will gain the most altitude per unit of time. We almost always utilize this speed as we climb the airplane for a normal trip cross-country or just out to the traffic pattern. Since this speed varies for every type of airplane, it would do us little good to be any more specific about it here.

There is one other climb speed that the beginning flier will learn. That speed is the best angle of climb. It also is a specific number and varies for every type of airplane. It is best used as an obstacle clearance speed, because the airplane will climb and gain the most altitude for every foot that it travels forward. To make this clearer, imagine two airplanes waiting at the end of the runway to take off. Both of these airplanes are identical in weight and horsepower. Plane A will take off and use the best *angle* of climb speed, hoping to clear a one hundred foot obstacle halfway down the runway. Plane B will take off and will use the best *rate* of climb speed hoping to clear the same obstacle. Only one of these airplanes has a good chance of making the flight beyond the obstacle. To make a long story short, call the ambulance for Plane B.

So there are the two main climb speeds a new pilot will learn before he gets that license in his hand. There are techniques for

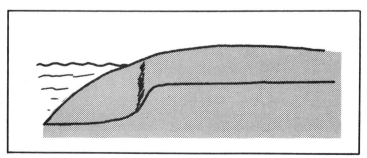

Fig. 3-4. This is the attitude for a normal climb.

accomplishing both climbs successfully, but the best rate of climb is what is used 99 times out of a hundred.

As we mentioned earlier, climbs are the biggest part of the first lesson. Only when we get to the following lessons does the student begin to readily identify climbing attitudes. First, there is the normal climb straight ahead. The pitch is controlled by bringing the cowling up to the horizon and watching (scanning) the airspeed indicator for the proper best rate of climb speed. When a turn is introduced during a climb, the references go astray. The largest part of the horizon disappears under the cowling. For most of us this becomes slightly disconcerting, because our main reference is almost gone. Only a remnant is left to use. Another problem that comes with the climbing turn is the addition of back pressure. When we bank the airplane, remember, a certain amount of lift is lost. Therefore, we need to increase the angle of attack or back pressure. The combination of these two factors make the climbing turn a little more difficult to master than most other maneuvers.

What can we do to make the climbing turn easier? There are several things. Foremost, we must use what outside reference there is. Granted, it won't be much. The way I have always done it is to use the so-called corner of the cowling. In the climbing turn, the corner pitches up and obstructs most of the rest of the horizon. In a right turn, for example, just a little of the horizon can be seen to the left of the corner of the cowling. A great deal more can be seen to the right of the corner. The object here is to raise that corner about three inches above the horizon on each side while maintaining a 30° constant bank. The airspeed will remain near constant and small pitch adjustments can be made with back pressure to maintain the best rate of climb airspeed.

When turning to the left, the process is the same. The references are reversed with the large piece of the horizon on the left side and the small piece on the right side.

Another technique that eases the mastery of the climbing turn is to use trim. Once again, I point to the fact that an airplane will lose lift in a turn whether it is climbing or descending. That loss of lift must be compensated for by the addition of back pressure and the increase in angle of attack. We have gone over this a great deal, but it is one critical area that can wind up lacking in a student's performance. To compensate for that loss we increase the back pressure. There is one other way that makes the maneuver much easier—trim. That trim wheel sitting down there between the seats is just begging to be the center of attention. If a student would learn to use that little wheel for every pitch change, the road to the Private License will be short and sweet.

We don't always use the trim wheel in every turn that we make. The reason behind this is the fact that not all turns are of the duration that would make physically holding the back pressure with the yoke an uncomfortable job. Climbing turns are a little different, though. The airplane is climbing and the airspeed is already low. When the airspeed is low in an airplane, a condition known as slow flight, the control inputs are normally a good deal larger than those when the airspeed is near cruise. This is the case in a climbing turn, hence, the reason we usually add trim for the turn. As we reach the 30° we add back pressure in a level turn. In a climbing turn we do the same thing. The exception this time is that it is usually just as easy to trim the plane instead of hold the back pressure. One good pull on the trim wheel will ordinarily bring the amount of back pressure that one has to hold in his hand to a tolerable level. A little fine tuning and a perfect climbing turn will be executed. Using the trim wheel in any flight situation will make you a smoother and more precise pilot.

Descents and Glides

I once heard a person asking my cousin about flying. My cousin had a great deal to do with my being a successful professional pilot and is a pro pilot of over 13,000 hours himself. After asking several questions about learning to fly, the person finally asked if he thought that flying was really safe. My cousin retorted, "My dear, they have never left anyone up there yet." Well, a wisecrack never so aptly stated Newton's Law. The old adage of what goes up must come down is extremely appropriate to our next subject: descents.

If straight and level and level or climbing turns are hard to master, the saving grace is a glide. The airplane will glide whenever power is reduced to idle. The plane will not fall to the ground if the engine stops. After all, the engine did not *fly* the airplane aloft, the wing did. The engine only pulled the wings aloft. So when we reduce power, the airplane simply begins to fly downhill.

Glides can let the airplane down slow or fast. If we were cruising at normal cruise power and pulled the throttle back to idle, a fairly rapid descent would begin. The reason for this lies in a fact that we have not yet discussed. Understand that this next fact holds true for all phases of flight but is especially noticeable in descents. An airplane will seek its trimmed speed. Every airplane is trimmed for a speed. In a climb, we trim the pressure off to achieve the best rate of climb speed. In cruise, we trim the airplane for straight and level and best indicated airspeed. If the power is changed after setting the trim, the airplane will try to go back to flying at that speed. In the case of a descent, the power will be reduced by some measure, and the nose will then fall below the horizon in search of its trimmed speed. Descents usually follow cruise flight, though not always. When a descent follows cruise, the airplane is already trimmed for a high speed. For example, a single-engine fixed gear plane may cruise at about 120 mph. If the power is reduced, the nose lowers and the airspeed falls off and then begins to increase. There may be several oscillations but eventually the airplane will settle into a constant descent or glide. The difference between a descent and glide is that glides are made with power off. A glide can be a descent, but a descent does not have to be a glide.

Now we know how a descent comes about, but what makes them easier than the other basic maneuvers? They are generally easier in that the airspeed that is maintained in most descents need not be very specific. For cruise descents to an airport after a cross-country jaunt, the only limit we need to watch out for is the red line (never exceed speed) on the airspeed indicator. In most cruise descents we just pull the power back a little and let the wings sing to the stars.

If we turn during a descent, lift is decreased as it is in all turns. It doesn't seem to affect us so much though, since the airplane is on a downward push anyhow. The thing we must monitor is the vertical speed indicator. In unpressurized, trainer-type aircraft, 500 feet per minute is the best rate of descent. A turn after the rate of descent is established will serve to increase the rate of descent. An increase of a hundred or so feet per minute may not be worth retrimming for, especially if the turn will be of short duration.

Maximum performance glides are another animal. These glides are performed as practice for an emergency landing or the real thing. In this case, a specific speed is necessary to keep the plane aloft for as long as possible. Possibly the best way to look at what a pilot is attempting to accomplish by performing a maximum performance glide is to define it this way: A maximum performance glide is one in which the airplane travels forward the greatest distance with the least altitude loss.

Let's pretend that we were flying between two points and lost the use of our engine enroute. We know that there is an airport just a few miles ahead. With the altitude we have on hand, we feel from experience that we can indeed glide to that airport and land normally. The important thing is that the pilot and airplane will have to give maximum performance for the landing to be assured. The airplane can only be configured by the brain, the pilot. The pilot has several things to do in this pressing situation. The most important, however, is that of getting the airplane into the maximum glide configuration. I was always taught to trim for one important speed. That speed is the best rate of climb speed. Remember, that speed gives the greatest gain for the amount of time it is held. When we use that speed for a glide, we reverse the role of that speed to give us the least amount of altitude loss per unit of time.

Now there are those who are so knowledgable and would tell you that the best glide speed is not V_y (best rate of climb) but that it is a separate speed altogether. This may in fact be true for most planes. The trouble is that this speed is generally not published. Moreover, it is usually within a couple of knots of V_y. So it comes down to this: Most pilots (including the majority of airline pilots) cannot hold an airspeed within a couple of knots. Also, since the speed is so near to V_y, why make it necessary to remember two speeds, one of which you may be fortunate never to have to use. One other thing—there is not an airspeed indicator around that may be that accurate. If there is one, it probably got that way by sheer accident.

What happens when we find the need to use our best glide speed? Well, there are several things that an instructor will run through such as checking mags and trying to restart and talking on the radio. None of those items fit into this discussion at this time. What *is* important is that we achieve best glide speed as quickly as possible to conserve every foot of altitude. Wouldn't it be something to fall inches short of the runway and blow a tire on the edge of the concrete just because we didn't trim for best glide soon enough? The lesson is evident in such a case. Thus, when we lose power on the engine, the first thing we should do is to trim for best glide speed.

Trimming for best glide speed is the same as trimming for any speed or desired attitude. The one thing that makes trimming for best glide speed different is the lack of a common attitude. Since engine failure could come at the beginning of a flight or near the end, the weight of the aircraft due to fuel burnoff may always be different. This would result in a different attitude. The best way I have found to teach students how to trim for best glide speed was the way I was taught. At the indication that the aircraft has lost complete power, trim the nose up (trim wheel back) with three good pulls of the trim wheel. As the aircraft slows down from its cruise speed, for example, the trim that we have adjusted will tend to hold the airspeed near the required value. Once the usual cycling of the nose has settled somewhat, the airspeed will assume a value near the best glide speed. After this has taken place, small changes in the trim can be made.

In searching for a landing field or spot in emergency circumstances, it may be mandatory to turn the plane to view the ground below. I believe this is necessary. On the other hand, we should strive to keep the banks as shallow as possible to avoid losing precious altitude. The important thing to remember is that we need to stay aloft as long as possible until a suitable landing spot has been found.

Another important aspect to glides or descents is how they apply to landing. All landings are preceded by a descent or glide. There is no exception. There is a trouble spot, though, in handling glides properly prior to landing. When maneuvering close to the ground, there is a tendency for most of us to pull the nose up, away from the ground. This is called being ground shy. It most obviously manifests itself during the landing flare with a student. It can be a big problem, nonetheless, while the airplane is still several hundred feet in the air prior to landing.

Once a glide is set up for a landing it is best to leave well enough alone. We should trim for landing according to certain speeds prescribed by the aircraft flight manual and our instructor. After this has been achieved, the trimming should be left alone except when the addition of flaps makes it necessary to retrim. The glide path of the aircraft, if started at the proper point, will almost take a plane automatically to the proper point. An instructor will show any student where that point is for the airplane he is flying.

The tendency for new fliers is to raise the nose if they feel the airplane will not reach the runway. This is the *wrong* thing to do. Let me explain. First, the raising of the nose slows the airspeed. This will in fact trade a little airspeed for altitude and the glide will appear to be extended. The trouble with this technique is that the loss of airspeed will inevitably result in a higher sink rate and the upshot is that the length of the glide will be *shortened*. There is only one way to adequately stretch a glide and remain safe. That is to add power. If power is not available, then we must be correct the first time we try it. A good hedge is to set oneself up for a landing a little high, for there are techniques to lose excess altitude and none to *gain* altitude when already trimmed for landing speed. We will discuss this technique a little later.

The Phugoid Cycle

What the heck is the *phugoid cycle*? Well, we have already alluded to it in the previous discussion. The tendency of an airplane to oscillate between a nose-high and a nose-low attitude after the power setting or trim setting has been changed is called the phugoid cycle. As we mentioned, an airplane will seek its trim speed—or shall we say the speed for which it is trimmed. Whenever it does that, the nose invariably overshoots, then undershoots, coming closer and closer to the correct attitude until at last the trimmed speed is reached. It is more like a coin twirling on a tabletop and making smaller and smaller vertical variations at the rim until it finally comes to rest on one of its faces.

Now what does this have to do with flying an airplane properly? That may be a good question. We could trim an airplane and let it oscillate until it finally settled down. Or we might reduce the power and let it porpoise around the sky looking for the proper airspeed. The trouble with either of these techniques is that it is very shoddy flying. It makes passengers uneasy and lets them know without a doubt what a terrible pilot you are. So we don't do things that way. What we *do* is fly the airplane positively. For example, if we reduce power to initiate a descent, we know that the airplane will have a nose-low attitude. What we will do then is to add forward pressure and some forward trim to establish the desired attitude. The tendency of the aircraft to oscillate will be negated by positive control inputs. By doing this we eliminate any uncomfortable side effects of changing an airplane's attitude.

Chapter 4

Simple Attitude Instrument Flying

The FAA has, for several years, favored the idea of integrated flight training. To the FAA, *integrated* flight training means the student will receive some instrument training in the maneuvers he is learning to do visually outside the window. The idea behind this comes from the accident statistics. An inordinate amount of aircraft accidents every year are the result of the pilot continuing into adverse weather conditions which demand more skill or equipment than were at hand. Although there are several ways to attack this problem (such as better weather recognition on the part of pilots), a higher level of preparedness to fly in instrument conditions is also an answer. Hence, most flight training contains some simple attitude instrument flying.

Attitude instrument flying is the technique for flying specific maneuvers using instrument references. Flying by instrument reference can be in some ways easier for the new student than to look outside for that elusive visual reference. The way that instrument flying is introduced will usually make the difference in how the student responds to the exercise. Actually, the whole process can be divided into parts. For instance, a standard rate turn can be taught to a student by a simple 1-2-3 method. It is this method for every maneuver we will discuss in the following pages of this chapter after we clarify how each flight instrument reacts to various maneuvers.

The Flight Instruments

In this section we will discuss what each flight instrument does to give the pilot information on the inflight attitude of his aircraft. There are only six flight instruments that we will discuss in this chapter. They are the airspeed indicator, attitude indicator, altimeter, turn-and-bank or turn coordinator, directional gyro and vertical speed indicator. The attitude indicator has gone through many name changes. It has been known as the gyro horizon, artificial horizon, and ADI for military buffs. (I believe that the term attitude indicator is most commonly in use today and is what I have settled on using in this book.) Now, let's examine each one and what to expect out of it in the way of information.

The Airspeed Indicator. The airspeed indicator is used just as a speedometer is used in a car (Fig. 4-1). There are certain speeds that we try to obtain and hold and there are speeds we endeavor never to exceed. What is not widely realized by most new pilots (or most old pilots, for that matter) is that the airspeed indicator's main function is that of an angle of attack indicator. We use the airspeed indicator to fly the correct speed to give us the best rate of climb speed, best angle of climb speed, etc. In essence, what

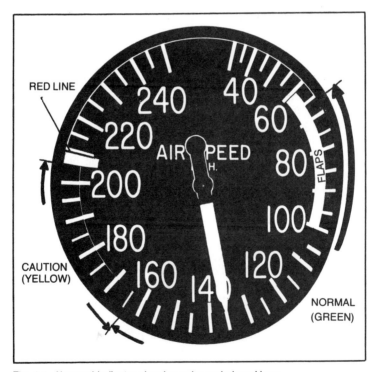

Fig. 4-1. Airspeed indicator showing color-coded markings.

we are doing by flying these speeds is positioning the wing for the most effective angle.

The airspeed indicator can also tell us at a glance whether the aircraft is ascending or descending. Obviously, if the airspeed is decreasing, the airplane is in a climbing attitude. The analogy is to that of a car going up a hill without using the accelerator any more than at the bottom of the hill. On the other hand, if the airspeed is increasing, the aircraft must have its nose in a lower than level position. It is the same as a car coasting down a hill; the car picks up speed.

It is not in the scope of this book to explain how each flight instrument works mechanically. There are plenty of books that do that already. However, there is one other important thing about the airspeed indicator. Airspeed indicators have color-coded arcs around the dial face. I think we should explain each one. First, there is the white arc. This is the flap range. If the airspeed indicator is within this arc, it is safe to operate the flaps either up or down. The bottom of the white arc is the calibrated speed at which the airplane is expected to stall with full flaps extended. The high end of the white arc is the maximum speed at which the flaps can be extended. Extension above this speed could possibly cause damage to the flaps and airframe.

Next, there is a green arc. The green arc is the normal operating range of the aircraft without flaps. Once again, the low end of the green arc is the calibrated speed at which the airplane is expected to stall without flaps extended. In comparing the white arc to the green arc, one can see the airplane will stall at a slower speed with flaps extended than in a "clean" airframe configuration. The top of the green arc is the maximum speed the airplane can be flown in turbulence.

The caution range or yellow arc is an area that is not used very much. I would like to point out however that it is perfectly permissible to operate an airplane in this "high speed" range as long as the air is smooth. The fact that turbulence is hard to predict, though, makes operation in this range a little dubious.

Finally, there is the short red arc. This is the never exceed speed. This does not mean that if one goes faster that the wings will fall off, although they may. What this line actually means is that the aircraft was never flown any faster than this during its certification process. Therefore, to go beyond this line puts one in the test pilot category.

I can attest that stepping over the line does not mean the automatic end of life in the cockpit or anywhere else. One spring afternoon, a student and I were out to check him out in a Cherokee 180. These airplanes are not placarded against doing spins. In fact, the spin entry speed is clearly marked above the glove box. The problem we had on this day was that I had never spun a Piper aircraft. In Cessna aircraft, to recover from a spin all that is required is to release the back pressure on the controls. The Piper aircraft, although we didn't know it at the time, requires a very generous push forward on the controls to break the stall during a spin.

Demonstrating the process, I pitched the 180 up to a high attitude. We got the usual Piper buffet and shiver. I kept the pressure back on the controls. The nose sliced through the horizon like a razor through a curtain. The gulf coast tidelands began to spin below us. After about a turn and a half I applied opposite rudder to the spin. She stopped spinning immediately. At the same time I released the back pressure on the controls and waited for the nose to start back up towards the horizon. *Nothing.* That's what happened. It's funny how fast one can break into a sweat and just go sick at the same time. The airspeed was building. The airplane was rock, just as if we had been thrown from the top of a tall building. The nose was headed straight down for some guy's duck blind. Completely shocked that the right things were not happening, I came to my senses. I gave a very hardy push on the controls. The stall broke but the nose was now pulling us through the beginnings of an outside loop. Whoops! We don't want to go that-a-way. Now, I could add back pressure and get some response. The rock had turned into a rocket. The wind was really whistling as we came through the pullout. The airspeed busted the red line by about 10 knots and the wings stayed on. To me, it was a wonder.

Looking over at the student, I got a picture of blissful ignorance to what had just taken place. "Man, that was fun. Let's do it again!" I looked at him through my fear-dilated eyes and said, "Let's not, the lesson is over."

The Attitude Indicator. The attitude indicator is the main information source during most instrument flights (Fig. 4-2). The reason for this is that it can give us pitch information. It can give us roll information just as well. The one thing it does not tell us about is yaw. However, on some airline models of the instrument, an inclinometer (the ball) is placed at the lower edge of the instrument so that all three axes are directly in front of the pilot. Unfortunately,

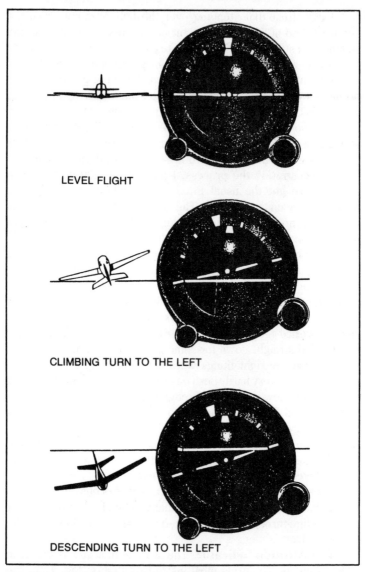

LEVEL FLIGHT

CLIMBING TURN TO THE LEFT

DESCENDING TURN TO THE LEFT

Fig. 4-2. Various indications on the attitude indicator.

you will probably learn to fly without this state-of-the-art instrument.

The accompanying illustration shows how to interpret the gyro horizon or attitude indicator. The illustration is, however, devoid of the 10, 20 and 30 degree marks that *do* appear on most modern instruments today. This instrument is quite reliable in most normal flight attitudes. It has no obvious quirks about its operation. Turn the airplane upside down, though, and we have a tumbled and useless piece of equipment.

The Altimeter. The altimeter is probably the most important instrument in the cockpit. This stands to reason when we think that it tells us how high we are in relation to sea level. If we have a chart that tells us what the ground elevation is below us, we can do simple subtraction and know if we are safely above it. It is the separation from the ground that is so important. A few pilots have tried to fly through terrain, but none have ever made it.

From the standpoint of using the altimeter as an information source for flying without visual reference, the exercise is simple. If the nose is in a climbing attitude, the altimeter will increase the altitude shown or indicated. The converse is true when it comes to descents. Altimeters change instantly with any variation of atmospheric pressure. Thus, it is necessary to set them often to the nearest reporting station near where we are flying. Also, since the altimeter responds rapidly to changes it is the most important instrument for maintaining level flight.

In instrument flying there are primary and secondary instruments. We will explain this fully in the coming pages. However, suffice it to say at this point that the altimeter is a primary instrument in straight and level flight and level turning flight. It plays this role because of its quick reaction to changes, making it easy for a pilot to fly precisely without visual reference. As far as quirks go, the altimeter has none as long as the proper setting appears in the Kollsman window at the right.

The Turn and Bank Indicator. The turn and bank indicator is probably the hardest instrument to understand (Fig. 4-3). The problem comes from the fact that it indicates a turn, yet it does not indicate how much bank is being put into the turn. The turn and bank is the indicator of the *rate* of turn. For example, the "doghouses," as they are called, are the marks of a standard rate turn. A standard rate turn by definition is a turn that covers 3 degrees per second. A standard rate turn will cause an airplane to completely reverse its direction in one minute because it takes one minute to turn 180° at the rate of 3° per second. To turn all the way around (do a 360°) takes two minutes.

There is an instrument which is a variation on the turn and bank which is often found in modern light aircraft. This is the turn coordinator. It has wings which dip in the same direction as the wings on the aircraft itself. The use of this instrument is sometimes misunderstood because it looks like the wings on the attitude gyro. New pilots tend to believe that it simply duplicates the roll function of the attitude gyro. The only thing that either instrument does is to indicate that a turn has begun in some direction. If the needle or the wing goes to the mark, a standard rate turn is in progress.

Standard rate turns are important for a couple of reasons. First, a standard rate turn in a relatively slow aircraft usually means the airplane will be banking less than 30°. In instrument flight, a medium bank or shallow bank is easier to handle because it tends to be more stable and there is less chance that it will induce vertigo. Secondly, a standard rate turn will get the pilot and his airplane headed away from bad weather and back the way he came in just one minute. This can be an important thing to remember if you are not actually instrument rated. A 180° turn is the best way to get back to better weather. And let me emphasize that the 180° turn should be performed at a standard rate.

At this point we should emphasize that the turn and bank is an important instrument for making a turn when using just instrument references. When flying on the gauges, it is customary and good operating procedure to make all turns at a standard rate due to reasons explained previously. The other reason is that radar controllers expect to see standard rate turns. If they are vectoring planes all over the place, they expect each aircraft to take up only so

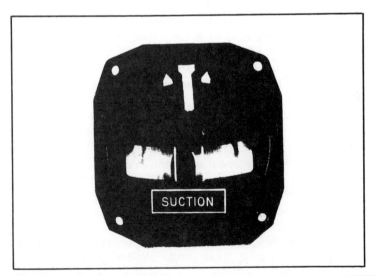

Fig. 4-3. Turn and bank indicator.

much space in a turn. If for example, a pilot who is only VFR rated is in trouble and is being given radar assistance, it would also be important for him to try to fit in for maximum safety.

The inclinometer is also on the face of the turn and bank or turn coordinator. We have already explained its function and under instrument conditions it performs the same job.

The Directional Gyro (DG). The directional gyro is also one of the basic instruments of the standard T arrangement on most modern flight panels. When set to the magnetic compass, it gives us minutes of reliable heading information. Most DGs are guaranteed to be accurate for about 15 minutes. Therefore, every 15 minutes or so, the pilot must check the heading against the magnetic compass. Ordinarily, the DG is set while the aircraft is in straight and level flight. Doing this minimizes the error in reading the magnetic compass.

The information that can be learned from the DG in flight is simple indeed. It takes no brain to realize that if the numbers or heading is changing, the aircraft must be turning. If the numbers are decreasing the airplane is turning to the left; the converse is true. This may not seem like much, but it can be very important in a situation where some of the other main flight instruments become inoperable for some reason.

The Vertical Speed Indicator (VSI). Last and probably least is the vertical speed indicator. Out of the six flight instruments on the panel, this is the one we could probably do without the easiest. The VSI gives us information about our rate of ascent or descent. Also, in simpler terms it will tell us if we are going up or down. The VSI is most probably used as a gauge to set the rate of descent for passenger comfort. The rate most often used is 500 feet per minute. This seems to be the rate at which ears can best adjust to pressure changes and yet let the airplane descend at a fairly good clip. The rate of ascent is less important because it is generally controlled by the power output of the engine and the effectiveness of the wing.

There are some important quirks about the vertical speed indicator. The foremost one is that the instrument is not instantaneous. There is usually a five second lag between initiating a descent and it showing up on the VSI. In some models of the instrument, the lag can be as long as nine seconds. As one can see, this would not lend itself readily to precision flight. Another anachronism of the instrument is that it is affected by turbulence. A good positive updraft could cause the needle to indicate a descent momentarily before registering the fact that the airplane is moving upward with the air. An equally exuberant downdraft will cause the reverse reaction.

As for actual information for the control of the airplane by instrument reference, the VSI has the least important role. Because of the inherent design problems of the instrument, it is usually used as a double check on the other flight instruments.

The 1-2-3 Method of Learning to Fly Instruments

I learned early in giving instrument flight instruction that the whole matter could be made easier for the student if the process was broken down to a few trite steps. Most of the normal flight maneuvers can be performed by following a pattern of three or four steps. So when we say the 1-2-3 method it may be the 1-2-3-4 or more method. The idea is to give a pattern of flow to the student so that his instrument scan is developed to cover the correct items for each maneuver.

What are the maneuvers that the private pilot student will have to learn to perform? As far as I know, there are no specific guidelines on this subject. It *is* safe to say that the basic flight maneuvers such as climbs, descents, straight and level and turns during each of the preceding are to be expected. One other thing is required, and that is the recovery from unusual attitudes. If you are not familiar with that term, it is an exercise whereby the instructor puts the airplane in an attitude that might result from spatial disorientation while you close your eyes and then gives the airplane back to you and lets you recover in a safe manner. We will discuss these in detail a bit later.

We have already discussed each of the flight instruments. Now we will attempt to use them in a logical system. Since each student's initial experience at controlling an aircraft by instrument reference usually begins with straight and level flight, that is where we will begin. Let me introduce the "hood." If you haven't heretofore heard of the dreaded device, let me tell you about it. The FAA calls a hood a "view limiting device." (It doesn't have to be a hood, but usually is.) The hood is a plastic visor which fits down over the forehead and has a bill that is long enough to annihilate the view of the horizon out the front windscreen.

After the plane has been leveled, the instructor will take the controls and allow you to put on the dreaded "hood." You are now out of touch with almost every reference that you have thus far been able to use. The most important thing to remember while flying instruments is that nothing has changed about the airplane. It still flies just as it always has. The control pressures are the same, the power settings are the same, everything flies the same! Some students get so flustered that they fail to assimilate all the information that they can still see. After all, you have only lost *one* source of information—the horizon. All other information remains fixed and usable before you (Figs. 4-4, 4-5).

With straight-and-level flight, the primary instrument reference is the attitude gyro. Everyone will eventually develop his own instrument scan. The scan cannot be taught, but it can be developed by the student on his own in a short amount of time. The important thing to do is to use the attitude gyro as if it is a hub of a wheel. For example, begin with the attitude gyro and then look at the altimeter. Then glance back at the attitude gyro. Then look at another instrument. You will soon develop your own scan.

I have found with examination of my own scan that most of my attention is divided between the attitude gyro, the altimeter, and the directional gyro. About every third or fourth circuit I check the airspeed. Indeterminately, I check the ball to make sure that the plane is still trimmed or that, due to some nervous tension, I have not begun to depress one of the rudder pedals unknowingly. With the hub and spoke method of scanning, you will soon become proficient in covering the instruments adequately.

Well, now we know which instruments to look at and in which order. The trouble is that you may not know just how to use what you are seeing. The attitude gyro is studied for any indications of a wing down or the nose not on the level. Incidentally, there is an adjustment at the bottom of every attitude indicator that will allow the pilot to position the airplane's wings in the indicator to a level position.

For fine adjustments in pitch attitude, we watch the attitude indicator. If the nose has strayed from level, however, there will be a corresponding change in the altitude, small though it may be. In cases such as this, it is easiest to adjust the altitude first by watching the altimeter and then go back to the attitude indicator to retain level.

By now, you may be wondering where is all this magic 1-2-3 stuff that idiot was talking about. Well friends, it doesn't really work with straight and level. Since all six flight instruments must be included in the scan, and some more often than the others, it is impossible to limit it to a memorizable order. That is why I said that

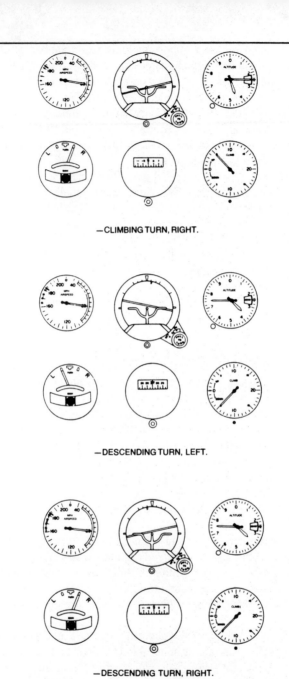

—CLIMBING TURN, RIGHT.

—DESCENDING TURN, LEFT.

—DESCENDING TURN, RIGHT.

Fig. 4-4. Various instrument indications for attitude instrument flight.

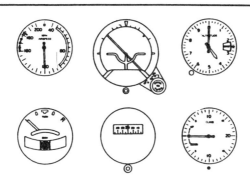

—STEEP, BANKED TURN, LEFT.

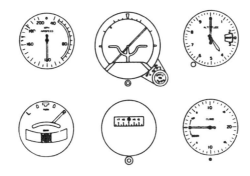

—STEEP, BANKED TURN, RIGHT.

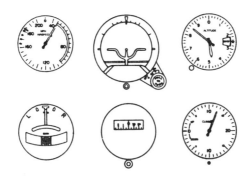

—APPROACH TO STALL.

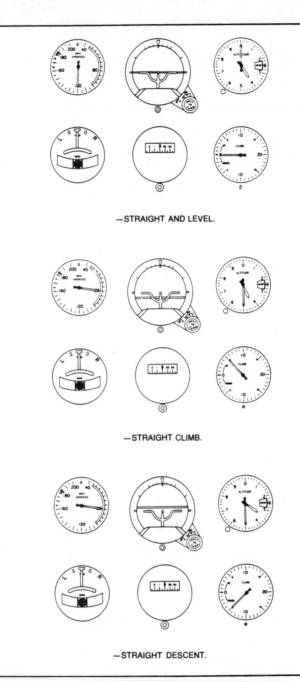

—STRAIGHT AND LEVEL.

—STRAIGHT CLIMB.

—STRAIGHT DESCENT.

Fig. 4-5. Some more indications for attitude instrument flight.

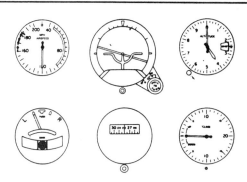

—STANDARD RATE, LEVEL TURN, LEFT.

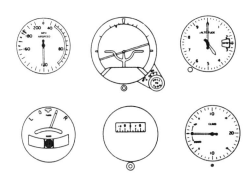

—STANDARD RATE, LEVEL TURN, RIGHT.

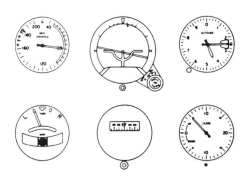

—CLIMBING TURN, LEFT.

everyone must develop his or her own scan. On the other hand, turns do take on an organized form and we will discuss them next.

Turns by the Number

In every turn there are several events that must happen in a preordained order. For example, the turn must be initiated, the increase in bank must be stopped at some point, the altitude must be maintained, and the heading indicator must be watched to know when to roll back level. The instrument references we have on hand do all that if we put them to work. So here is the sequence for beginning a turn.

1. Watch the attitude indicator to initiate a bank in the proper direction.
2. Observe the turn and bank indicator for standard rate and coordination.
3. Watch the altimeter after the bank is established as it becomes the most important (primary reference) instrument to watch.

This simple hop, skip and jump routine will cover any turn entry. There is a process for rolling out of a turn as well. First, we have to know some things about the turn we are in. It is supposed to be standard rate, but that can mean the bank might be anything, depending on the airspeed we are flying. We will assume that the bank is something less than 30°. In a light single-engine trainer, it generally is about 18°. Once we know the bank angle we need to use the rule of thumb for rolling out of turns. We should lead the heading that we need to roll out on by half the angle of bank. If the bank is 18° we should begin to roll out of our turn about 9° ahead of time. We could make it 10° just to be easy on ourselves.

So now here are the steps for rolling out of a turn on instruments.

1. Begin roll out one-half the angle of bank ahead of desired heading.
2. Roll out at an even rate watching the attitude indicator.
3. Upon reaching level, check the altimeter.
4. Double check the heading indicator.

This is the main flow for rolling out. Somewhere during the process, you might want to check the ball to stay coordinated. However, if most of your turns are well coordinated as a matter of habit and good instruction it is not that important. As we gain experience in the air we can feel when the airplane is uncoordinated and check then. For the beginner, it wouldn't hurt a thing to check.

Setting Up Climbs

As we mentioned earlier in the book, all maneuvers that change pitch should use the "Pitch, Power, Trim" formula. Although this has nothing to do with instrument references, it is important to the flying of the aircraft nonetheless. We should perform these functions almost unconsciously after a while without respect to the instrument chain of events. The sequence for establishing a climb is as follows:

1. Pitch the aircraft by using the attitude indicator; about a 5° pitch up is all that is required initially.
2. Establish the proper airspeed on the airspeed indicator and trim the airplane at that time.
3. Monitor the altimeter.
4. Monitor the heading indicator. Be ready for the heading to wander to the left due to high torque.

During the entire climb, the altimeter should be the primary reference. This follows the procedure that we use for climbs when we can see outside, so we should be quite comfortable with this part of instrument flying.

Descents are not unlike climbs in the instrument environment. In fact, the same sequence of items is on the agenda for establishing a descent. There are a couple of obvious differences such as the airplane is going down instead of up. Here is the sequence for establishing a descent in the instrument environment:

1. Bring the aircraft to the desired pitch attitude by using the attitude indicator. About one horizon bar width is all that is needed for a 500 fpm descent.
2. Establish the proper airspeed on the airspeed indicator and trim. Usually this is not as critical as it is in climb.
3. Monitor the altimeter.
4. Monitor the heading indicator.

As you can see, the steps are almost the same. In the descent, however, we can go just about as fast as the limitations on the airplane. In a plane with a fixed pitch prop, we must also monitor the engine rpm.

Unusual or Critical Attitudes

Unusual attitudes are those attitudes that are not common to ordinary flight operations. In other words, they are of almost an aerobatic nature. In fact, by the FAA's definition they would be aerobatic. This is not to scare you away from flying if you think you will be doing aerobatics and that is not for you. Your flight instructor will not be turning the plane upside down or anything close to it. You might be flying through some steep turns and dives where the nose dips beyond 30° down, but those are not that bad.

The importance of critical attitudes cannot be overstressed. The reason we go through them is to gain proficiency in recovering the airplane after spatial disorientation by believing our instruments and not our sense of kinesthesia. The one thing that the pilot faced with an unusual attitude must do is to *believe the instruments*. After flying through the preliminary buildup that the instructor will put you through, it is normal to think the airplane is doing something other than what the instruments are telling you it is doing. This feeling is kinesthesia, your sense of motion. When we have no visual references to tell us what is up and down and which is right or left, we can become mixed up. A study of the instruments will tell us just what is going on. After that it becomes an important part of the exercise to recover the plane in a safe manner.

The recovery from an unusual attitude can be learned and memorized. In fact, it can be reduced to a 1-2-3 method like all instrument flying.

The most important items in a recovery from an unusual attitude are in their respective order, *airspeed, attitude, altitude*. If you are wondering about this sequence and its logic, listen to this. One would think that since we are dealing with a critical attitude that the attitude would be the most important thing to correct at first. Such is not the case. The important thing is the airspeed. Let's look at the reasons.

A critical attitude is a situation bordering on the dangerous. It will become very dangerous if the trend is not reversed. First. not all critical attitudes are nose-low. Some are nose-high. So we may be dealing with two distinct possibilities. The airspeed may be bleeding off at such a rate as to bring the onset of a stall. In such a situation, an unaware pilot could possibly stall and spin. The opposite end of the ruler holds the high airspeed card. A high airspeed could possibly result in tearing the wings off if the aircraft is not slowed. Therefore, airspeed is the item that must be arrested at once in an unusual attitude.

The next item to attend to is the attitude. We can glance at the attitude indicator and at once know whether we are in a right or left turn. Also, noticing the airspeed (whether it is low and going lower or high and getting higher) will confirm whether the indication on the attitude indicator is correct (the other possibility being that the attitude indicator has tumbled during the entry into the attitude). In a nose-low attitude, it is most important to level the wings first or prior to raising the nose back to the horizon line. The reason for this goes back to Chapter 1 where we discussed load factors on airplanes in a turn. So remember, if the nose is low and airspeed is high, level the wings before pulling up to level off.

Whenever the airspeed is low or we have a nose-high attitude, the recovery process is not any less dangerous or hurried. It must be handled expeditiously and properly. The principal difference between the two recovery techniques is that the wings don't have to be leveled immediately. In a low airspeed condition, we must lower the nose at once to eliminate the possibility of a stall. Since the airspeed is low, there is no likelihood of overstressing the wings if we lower the nose in a normal fashion. Simultaneously, we can level the wings.

The other reference we must pay attention to is the altimeter. Most of us pilots keep a running tab on the elevation of the terrain beneath us. I don't mean to know every hill and mountain, but rather the average terrain elevations. For example, we are flying over farm country that has some hills but most of the airports in the area have field elevations of around 1200 feet. If we got into hot water and the altimeter was unwinding, we would know just about how quickly we must affect a recovery. (Although I must say that recoveries should always be done *at once*.)

The altimeter also tells us something else besides how high we are flying. A glance at the altimeter will confirm what we should already know. The aircraft is either climbing or descending. I should point out however that it is not always that important in speeding us to a recovery. The airspeed and attitude indicator usually provide enough information in a hurry to take care of the situation.

We should now turn our attention towards the recovery technique. Once we have the information as to whether we are in a climbing or descending turn at the operating limits of the airplane, we must take action. The first action we should take is to decide whether we are nose-high or nose-low. Next, we should note the airspeed. After that we have the information to act in the proper fashion, which is to adjust the power. If the airspeed is low or falling, the action to take is *max power, lower the nose*. If the airspeed is high and increasing we should do this: *power to idle, level the wings and raise the nose*. These responses will save your life. I must say one thing here, though. If you are fairly competent at attitude instrument flying and practice it every once in a while, along with avoiding weather which is beyond your experience level, a critical attitude will never crop up.

Now, let's outline the proper 1-2-3 method for critical attitude recovery.

1. Notice the airspeed. Is it increasing or decreasing?
2. Note the attitude indicator for a nose high or nose low attitude.
3. Is the altimeter winding up or winding down? This is the confirmation step for the other two references.

After we have realized our predicament by the instrument references we will be faced with the proper recovery reaction. In a nutshell, here they are again.

1. Airspeed high and increasing demands that we pull the power back to idle, level the wings, and raise the nose.
2. If airspeed is low and decreasing, we should apply maximum power and lower the nose to level. The wings may be leveled at any time.

What to Expect from Your Instrument Training

We haven't covered a tenth of the curriculum of a regular instrument pilot's rating. In your training for Private Pilot a few more exercises such as timed turns and partial panel exercises will be covered by your instructor. Even so, no pilot with this limited amount of training will be ready to go out and fly through clouds. Believe it or not, very few newly instrument-rated pilots are actually safe to go out and fly through clouds. It takes experience, which is built up over a period of palm-sweating flight hours. With the knowledge that the average private student will accrue, he is only a *little* more prepared to save his own life should he/she wander innocently into a cloud. To think that you are is to ask for trouble. The best hedge the Private Pilot that is non-instrument-rated can have is a better working knowledge of the weather so he can discern when it is time to turn back to safety or to land right where he is.

Chapter 5

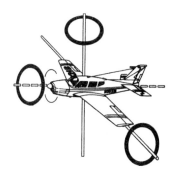

The Stall Series

Before we can go out and fly a great deal, and certainly before one can solo an airplane, a little greater margin of safety must be built into any student's repertoire. Flying an airplane of today's modern vintage is very safe in itself. The problems come from unsuspecting or unaware pilots getting themselves into situations that the airplane cannot handle. Of course, we refer to stalls.

Late in Chapter 1 we discussed how a wing stalls. The key here is *wing*. If anyone ever says to you that an airplane stalls because the engine quits or that a stall means the engine quit, kindly set them straight. *Wings* stall; engines lose power or quit. They do not stall as if we can pull off to the side of the road and look under the hood. The news media quite often says this when they present a story on a local airplane crash and I almost pull my hair out when I hear it. You can go a long way in educating others who are non-fliers if you would just explain this simple difference when it crops up in your flying future. They don't need a course in *how* a wing stalls, just tell them it was the wing and not the engine.

Well, enough about my pet peeve. Actually, though, stalls are a necessary part of any Private Pilot curriculum, Stalls can develop into situations where the pilot can lose altitude in a hurry. During some flying maneuvers (such as ground reference maneuvers) that we will discuss in the next chapter, or during landings, there is not much altitude on hand. In fact, an overwhelming number of accidents each and every year occur in the course of landing. Improper pilot response is generally the cause.

We will practice stalls aloft at altitudes that are higher than 2500 feet above the ground. For the average student and instructor this is a good safe margin for error and recovery. Believe me, most instructors are capable of regaining control in a second or two if a student screws up the whole deal. This should be a comfort to those who feel that learning stalls puts them out on the ragged edge.

To begin with, there will be four basic stalls that you, the student, will need to learn. The first one will be done with power off in most cases, and is easy to learn and recover from correctly. Unfortunately, this stall procedure is ordinarily used as a confidence builder and will not be required of the student on the check ride for Private Pilot. We will discuss all four of the main stall series here in this chapter and explain how each one relates to other flying situations. If we can understand that stalls relate to normal flying maneuvers, we can then stay away from the sloppy techniques that would bring on an unanticipated stall.

I cannot stress this point enough. Some students come away from a training program saying "I won't ever accidently stall an airplane because you have to work too hard at it to get it to stall." This may be true to a point. To perform the stall series as if we were on a check ride calls for a little work. These students have missed the idea behind learning stalls. The idea is to recognize the onset of a stall and not all the rigamarole that goes into causing one. Every stall type that a student learns has its basis in one of the regular flight maneuvers. That is how they will be presented here. If your instructor does not do it the same way in the airplane, you are being shortchanged in your learning process. The stalls will be outlined here, and if you are not instructed to do them in a manner that relates to an actual flight condition, you can always practice them that way on your own.

The Power-Off Stall

Getting acquainted with stalls is usually a nerve-wracking experience. There is no reason that it should be. The airplanes of today are controllable through the entire stall. It is just the idea that the wing quits flying that makes most of us uneasy. Believe me right now that the power-off stall without flaps is hardly noticeable from regular slow flight. In fact, slow flight is a great way to get acquainted with the handling characteristics of an aircraft nearing the stall regime.

Slow flight simply means flying the airplane slow. It also means, obviously, that we will fly the airplane at reduced power and at slower airspeeds. Do you recall the green arc on the airspeed indicator? Well, about five knots or miles per hour above the bottom of the green arc is where we normally endeavor to hold the airspeed during the slow flight exercise. Also, please bear in mind that slow flight can be done with flaps up, down, or anywhere in between.

Bringing the airplane into a slow flying condition, we must first reduce the power on the engine. As this is the first time that you will have pulled the power off in flight, you may want to do it gradually. The flight instructor will probably have shown you the entry to this maneuver before you try it. However, we will discuss it here as if it is the first time you try it. An old wise FAA flight inspector once told me that the best way to slow a plane down is to grossly underpower it. In other words, we know we will use a little power to keep it flying but not much. It would be most expeditious, then, to pull the power completely off to slow the airplane down. The converse of this rule is also true. If you need more power, add more power than you need if you have it. Then reduce the power to what you *do* need.

Let's say that we have pulled the power back now almost to idle. One of two things is going to happen—the nose will fall below the horizon and the airplane will descend in an effort to find its trimmed speed, or the pilot will trim the nose up and the airspeed will begin to drop. This is the correct response and the proper one to bring the aircraft into slow flight. As the airspeed comes down to the bottom of the green arc, add a little power. At this point we must begin to control the airplane by a very important rule. *The power will control the altitude and pitch will control the airspeed.* Make no mistake, pitch is the varying pressure on the control yoke. At this point, the airplane's controls will feel very sloppy and the plane feels very wobbly and unstable. Slow flight is getting the student ready for two things: one, how the airplane will feel before the power-off stall and two, how it will feel during landing.

Expanding on the rule above, we must constantly watch the airspeed indicator and the altimeter as well as watch outside to keep the wings level. The reason we observe this rule is directly related to the aircraft's low airspeed. If we tried to control the airplane's altitude with the pitch control, we could increase the angle of attack beyond the critical angle and the airplane would never gain that much altitude, having stalled first. On the other hand, we can make small airspeed adjustments by varying the pitch a little. If we need more airspeed, we release some of the back pressure. If we need less airspeed, the ideal thing to do is to increase back pressure. Slow flight is easy as long as the changes are small and remain small.

For a moment, let's imagine that the airspeed has increased by 10 knots above what is desired for a good slow flight demonstration. At the same time the altitude has remained fairly constant. The reason behind this problem is that the aircraft has excessive power.

Now, you say "I thought that power controls altitude. And if it does, and the airplane has remained at the same altitude, then there must not be excessive power, right?" Not exactly. In fact, the reason that the airplane has gained airspeed is due to the excess power. Now, if we were to enforce our rule and controlled the airspeed with the pitch control, it would become obvious. We could expect that the airspeed would bleed off and at the same time the altitude would increase a little. The only response at that time would be to reduce the power and bring the altitude of the aircraft back in line.

Turns in slow flight take a great deal of finesse. Typically, the new student will try to use much too much rudder. It is a paradox of sorts that after putting so much effort into learning to use the rudder properly for normal turns, in slow flight the rudder is hardly needed at all. This will be true when we begin power-off stalls; not much rudder is necessary.

One other area about slow flight we should discuss before we move on is the use of flaps. Whenever we add flaps we reduce the stalling speed of the wing. When lowering the flaps during a flight, it is normal to feel a pitch change taking place. Usually, we must adjust the trim to compensate. If we continue to fly the airplane slow as we have been, we will now have to adjust the airspeed down towards the bottom of the white arc on the airspeed indicator. The controls will continue to feel sloppy and not much change will be noticeable in the lateral axis.

One way to tell if the airplane is flying about five knots above the stall is by listening for the stall warning horn. It is engineered to come on about four or five knots above the stall. If we are flying the airplane at the ragged edge of performance, the stall horn should sound intermittently. I have to admit that when I was learning to fly, it was a bit disconcerting thinking that the airplane might stall and I wouldn't be ready for it. On the other hand, it was good news that I was flying the airplane as the instructor wanted.

Power-off stalls are nothing more than slow flight carried to the extreme. They are docile, easy to control, and not scary at all. Let's talk about doing one and the recovery procedure. First, in Cessna aircraft we must pull the carburetor heat knob to full warm. In Piper aircraft we don't apply carburetor heat. Then the throttle is retarded slowly so as not to make the engine backfire. As the airplane begins to slow down, we must trim the nose back a little at a time. This makes it easier for us to keep the nose in the correct position. What is the correct position? It varies from instructor to instructor. However, if we put the cowling even with the horizon and hold it there, everything should work out perfectly. The thing we should bear in mind as the airplane begins to slow down is that the plane is entering the slow flight regime. Therefore, we can expect the controls to become sloppy, the outside air noise to become less, and the airspeed indicator to plummet. What we do not know to expect is how the stall will feel or exactly how the airplane will react.

As you approach the slow flight regime, the instructor will be telling you to increase the back pressure. Keep pulling back and this will bring on the stall more quickly. This is actually what you (the pilot) wants, because it lessens the amount of time to wait for the stall to take place. Now, we should be very close to the stall. The controls will be nearly back to your chest and very hard to hold back. The stall is at hand. A little shudder and the nose of the aircraft will begin to lower by itself. It doesn't lower much—only about what is normal for a normal descent. At this time, fully release the back pressure you have been holding and smoothly feed in power. The addition of power should be quick and positive, yet not so quick that it chokes the engine and causes it to quit altogether. In other words, don't jam on the power. As soon as all this has been done, the airplane will begin to pitch upward, ready to fly. In fact, one may have to "untrim" the plane or else the nose will pitch up too high.

Most manufacturers have built the critical angle of attack into their airplanes to be about 18 degrees. The airplane will fly up until that angle is reached and will begin flying again after a stall as soon as the airplane is relaxed to or below that angle. For example, let's assume that we just performed the power-off stall described above. If the airplane stalled at exactly 19 degrees, then as soon as we relieved the angle of attack back to 18 degrees, the airplane began to fly once more. We put an end to the stall almost immediately. If you are worried about how much altitude the airplane loses, let me say this: It will not lose enough altitude or lose it quickly enough for you to even feel it. A good average try at a power-off stall on the first attempt usually shows an altitude loss of less than 100 feet.

After the first one, try several more. They will become fun and familiar and build your confidence in your ability as a pilot. Some instructors will ask the student to throw in a shallow bank while doing the approach to the stall. This won't complicate matters very much. In this case it is a great deal like critical attitudes under the hood. As the airplane stalls, roll the wings level to eliminate any further aggravation of the stall. The one caution here is to monitor the ball as you roll the wings level. The turn must be coordinated or adverse yaw from the ailerons will cause the possibility of a spin. Don't let this scare you because it is difficult to spin intentionally from a power-off stall, but you should know the possibilities.

The Takeoff and Departure Stall

One of the stalls that will be required on the flight check for the Private Pilot rating is the takeoff and departure stall. Ordinarily, this will be the next stall that your instructor will teach you. It is one of the harder ones to do properly due to the fact that it requires full power. Full power in this case causes an unusual amount of torque and left turning tendencies. That leads to standing on the right rudder, similar to takeoff.

On the other side of the coin, the stall is easier to recover to normal flight. The reason is because full power is used to recover from *all* stall situations. Because we are already at full power in doing a takeoff and departure stall, we have only to lower the nose and fly the airplane out of the stall.

In the power-off stall section, we did not talk about how it applies to the everyday flying situation. It does in a way, though. A power-off stall type of situation could occur during a landing without flaps (a configuration that is called "clean").

In this section we will study how the takeoff and departure stall does apply to a real-life situation—takeoff and departure. To set the stall up we must first slow the airplane from cruise airspeed to takeoff rotation speed. Let's say that speed will be 65 knots. To slow down from cruise airspeed, we need to underpower the aircraft. The best way to do this is to pull the power to flight idle. The airspeed will begin to bleed off and the usual retrimming of the airplane is necessary. At the point that the airspeed reaches 65 knots we will apply full power. (Although we are in the air, imagine that we are rolling down the runway and are pulling abruptly back on the controls.) The aircraft will fly away from the ground, which is evidenced by an increase on the altimeter. We have become airborne, but the airspeed is bleeding off rapidly. What is actually happening in the plane is that the nose has pitched up about 25 degrees and the sky is all that is visible straight ahead.

In the actual flight situation, that of doing the stall, we need to add quite a bit of right rudder to keep the ball centered and the airplane coordinated. This is extremely important because if the airplane stalls in an uncoordinated state, an entry into a spin is a real probability. Meanwhile, back to the simulated situation. The airspeed is approaching that of the stall and we are only a couple of hundred feet above the ground. In a struggle to clear an object ahead and in our flight path we continue to pull back on the controls. Suddenly, the wing stops flying and the nose drops below the horizon. With full power already applied, all we have to do is to wait for the wing to begin to fly again. It will do this quickly and it is a good thing too, because the ground is only about a hundred feet below us.

In the real-life cockpit situation, the response is the same. We pull back on the controls to induce the stall. The stall is more abrupt than the power-off variety. (The first one I attempted took my breath away.) Now, all that is left to do is to release the back pressure we have been holding. The wing will begin to fly almost at once. Once it does, we should bring the nose back to the horizon in an effort to recover the airplane with the least amount of altitude loss.

Performing the stall is really very simple. It will be demonstrated by an instructor before you will be expected to do one. The one thing to remember about takeoff and departure stalls is that they represent a stall during actual takeoff and departure. I can't stress this enough. Stalls are not something that we must learn in order to pass the flight check. They are something we must learn to keep us alive in the event the plane stalls during an actual flight, whether the reason be a gusty wind shear or misuse of the controls.

The departure stall is just an extension of the straight ahead takeoff stall we just discussed. Ordinarily, departure stalls mean the stall is done with a shallow turn. Shallow turns are used to keep from aggravating the condition. Typically, departure stalls may dip a wing a slight amount at the time of the stall. It is even more important to keep the ball centered during the turn and stall.

The departure simulates a stall at a certain time of an ordinary flight. Do you remember the turn we must make after takeoff about 400 feet in the air? Well, this is where a departure stall might occur after takeoff. Keep in mind that if things are done normally every flight we should never see a stall at such an inconvenient altitude. This brings up another point: in the simulation of a departure stall, we actually have more altitude to recover the aircraft than with a takeoff stall. Nonetheless, we need to recover the plane in the least amount of altitude loss.

Approach to Landing Stalls

The name of this type of stall describes how the stall should be simulated. At an altitude of at least 2500 feet above the ground we begin the process of setting up the stall. As we mentioned, all stalls should be done at this altitude or higher to allow a safe margin for learning.

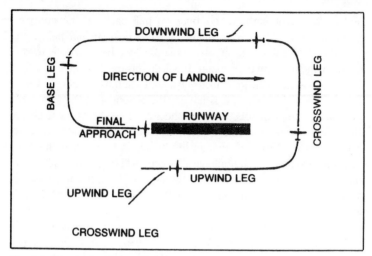

Fig. 5-1. Standard traffic pattern components.

Approach to landing stalls are customarily done with the flaps fully extended. Although you will find that not all landings are done with full flaps, an approach to landing for the flight check will be done in that configuration. The use of flaps in performing the stall exercise will make the airplane stall at a much lower speed. The stall will set in quite sluggishly and the nose will pitch down a little sharper than it will in a takeoff and departure stall. The good news, however, is that the stall is very gentle in any modern plane and the pitch over is what could be termed almost slow.

The setup for the stall is exactly like flying a traffic pattern for landing (Fig. 5-1). We will fly a downwind, base, and final leg. Note that not all instructors will teach this way. This instructor feels strongly, however, that relating a type of stall to a real flight experience puts the learning of the stall in the proper perspective. Since we fly downwind and base legs before turning final, the stall exercise should be set up in a like manner.

If it has never been called to your attention, be advised that the turn from base to final approach is the most dangerous turn in all of aviation. This turn has gained a bad reputation for all sorts of reasons, but the main one is that pilots pull the nose up and try to stretch the glide at that point. If they try to stretch it too much, the plane will stall and possibly spin due to the turn and uncoordinated flight controls. With the ground in close proximity, many a pilot has bit the bag.

Since we haven't yet talked about the landing procedure, take note here that this is in effect how it is done. All the steps to a normal landing through the turn from base to final are here with one notable exception, and that exception is the fact that full flaps will be selected before the final turn begins. Ordinarily, full flaps are selected after the aircraft is established on final approach and can glide to the runway in the event of complete power loss.

As with all stalls we must begin with a power reduction. That power reduction ordinarily comes when the airplane is abeam (opposite) the approach end of the runway or "threshold," as it is known. The type of aircraft one is flying will determine whether or not the carburetor heat is pulled full warm prior to the power reduction. After the power is reduced to flight idle, the airplane must be retrimmed. In almost every lightplane flying, three good, strong pulls back on the trim wheel will bring the airplane to its best glide speed. When the airspeed is within the white arc on the airspeed indicator we can then lower our first increment of flaps. This first increment is normally between 10° to 12° and is known as the approach flap setting. The airplane can be further slowed with these flaps dragging in the wind. At this time we should abandon our best glide speed and let the airspeed bleed back in order to produce the stall a bit later.

Once we have achieved the approach flap setting, we are free to turn from our downwind heading to our base leg heading. This will usually be a turn of 90° to the left. Picking a road or section line below as a simulated runway will help in keeping the landing pattern realistic. The airspeed will continue to drop and the air noise will be reduced. We can now add intermediate flaps and finally full flaps. Incidentally, full flaps will be between 37.5° to about 40° on most aircraft large or small.

Now that we have full flaps, we begin our turn to final. All the time we should be trimming the nose up, as holding the yoke back with full flaps takes a strong arm. Once again, use the road below as our make-believe runway. If we pull the cowling up to a level attitude or right on the horizon, the stall will not be long in coming. It is important for simulation's sake to stall the aircraft before it completes the turn to final approach. The stall will come directly after the stall warning horn sounds. Holding the nose up at this point is impossible. It will dip and we must release all back pressure as with all stalls. Our next reaction is to push in the carburetor heat, if it is selected, and then full power. At this time a great deal of forward pressure must be held against the controls. We must now compensate for all the trim we have rolled in and the natural tendency of the nose to rise when full flaps and full power work in tandem. If we do not hold this forward pressure firmly, we run the risk of the nose rising—and consequently, the angle of attack. The net result would be a secondary stall.

After the initial recovery reactions, such as releasing back pressure and adding power, we are faced with another problem. We have all those flaps hanging out there, working against us. It is common knowledge that after about 20° of flaps, any more that are added will be ultimately nothing but drag. If you are reading between the lines, then you have already figured out that we need to reduce the amount of flaps to at least 20° to expedite recovery. A word of warning in advance. This is the part of the procedure that most students fail to carry through. Most everyone will do the automatics like getting the nose down and power on. The heat of the moment, nevertheless, seems to obscure what is only a little less than obvious. Thus, this part of the exercise may need to be redone several times.

There we are, all those flaps hanging out and struggling to gain some airspeed. Popping the flaps back to at least 20° will have an instant effect on the airspeed—it will increase. At this point we should concentrate on keeping the nose level and not worry about climbing away from the ground. The remainder of the flaps should be "milked" off slowly. The reason for this is that the last 20° of flaps are mostly lift and not drag. To retract them too quickly will cause the aircraft to lose lift and possibly sink into another stall. The answer then is to get rid of the flaps slowly. As each successive increment of flaps is retracted, the airspeed will increase and soon the aircraft will be able to climb again out of the proximity of the ground.

Probably, the most important idea to remember in recovering from an approach to landing stall is to keep the nose level. The nose should be lowered only to level; if it goes below the horizon of its own accord, we should raise it to a level attitude as quickly as possible without inducing a secondary stall. The importance of this action is to keep us from losing precious altitude since we are very close to the ground. Also, approach to landing stalls tend to be a little "deeper" compared to power-off or takeoff and departure stalls.

On all check rides for a rating it is important to begin a climb after the recovery of every stall. At least climb to the original altitude at which the stall entry process began. The practical test guide calls for a climb of 300 feet. In all the check rides that I have taken I have never seen an examiner that didn't want to quit climbing before reaching that 300-foot level. Usually, they want to get on to the next maneuver.

To reiterate: the basic idea behind learning approach to landing stalls is to recognize where the common pitfalls to flying lie. Two things are important in flying a normal approach to a runway. One, the airspeed should be monitored closely and kept at the proper value. Two, glides should not be extended by trading airspeed for altitude. If you need to extend a glide, add power.

Also, one final note in doing approach to landing stalls: Don't worry about losing a little altitude while flying the pattern. We always lose altitude going around the landing pattern after we are abeam the landing threshold. Once the turn from base to final is begun, that is where the altitude should be held to simulate stretching a glide.

If the instructor of your choice does not teach the landing pattern approach to the stall, then I urge you to try it at least once. It will give you the proper perspective that is needed to learn why we do stalls at all.

Accelerated Maneuver Stalls

The term *accelerated maneuver stall* is a very important one. Within the definition lies the reason we learn to do these stalls. It may take a little reading between the lines, but the reason is there, sure enough. An accelerated stall results from the aircraft being induced to stall at a much higher than normal airspeed.

If we go back to Chapter 1, there was a section there on load factors. The accompanying chart showed that for a 60° bank in a level turn, the wings had to support twice the weight of the aircraft in level flight. Due to this fact, the wing will also stall at a higher speed. The airplane will not stall in a well coordinated and properly performed bank of 45°, 60° or even a 90°. However, in an abrupt use of the controls during one of the above banks, a stall will set in rather quickly. In performing an accelerated maneuver stall we essentially induce the stall by a quick action on the controls. It is not what one would call a true abrupt movement of the controls. Nevertheless, an abrupt movement will give you a stall (and the surprise of your life, if it was unintentional). For that reason, we must be prepared to know how to handle the accelerated stall.

To relate an accelerated maneuver stall to a real situation is not that easy. Other than steep turns, there is only one place where this stall might develop and that is during the turns in the landing traffic pattern. So, let's examine a proper execution of a steep turn so that we won't get ourselves in a dangerous situation.

For the Private Pilot rating, steep turns of 45° and of 360° duration are required for the flight check. This seemed to me, as I learned my Private Pilot curriculum, to be the most fun of all the maneuvers. Proper entry into a steep turn (or *any* maneuver, for that matter) begins with a clearing turn and a glance into the direction of the maneuver. Next, we begin to roll in the bank. A good steady rate of roll is required or else the airplane will have nearly reversed directions before the maneuver is established. We will stop the bank at 45°, of course. This is in the wide space on the attitude indicator halfway between the 30° mark and the 60° mark. There is no 45° mark on any attitude indicator that I have ever seen (Fig. 5-2).

The most important item in executing a steep turn is the maintenance of altitude. This begins at 30° of bank. At this point we must increase the back pressure substantially to maintain the altitude we have selected. We can do this in two ways, but a combination of both is best. First, select an outside visual reference. Due to the loss of lift in the vertical direction, a noticeable pitch-up of the nose will be required to maintain altitude. In a Cessna 152 for example, the horizon will cut the left corner of the cowling in a right turn of 45° bank about two inches below the corner.

The other way we can keep tabs is by instrument reference. Just as in medium-banked turns, the altimeter becomes the primary source of pitch information. We can use the attitude indicator, but due to the G force imposed by the steep bank, the unit will precess somewhat. Also, as minor corrections are made, the load factor or G force will vary somewhat, which causes the attitude indicator to vary in its pitch indication.

Once we are established in the steep turn, it is not easy for the beginner to maintain altitude (Fig. 5-3). The nose will hunt up or

Fig. 5-2. Instrument indications for a level 45° banked turn.

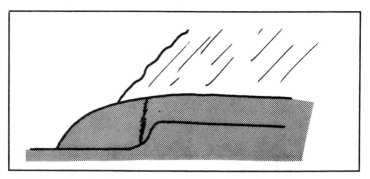
Fig. 5-3. Outside visual reference for a right-hand 45° banked turn.

down. Your first attempt will probably not be very well done. As a result, it may be necessary to recover from whatever mess you have gotten yourself into. It is not unusual for a pilot to find the bank steepening up. First, we should understand that it will take a little aileron pressure against the turn to maintain the bank. In addition, whenever the bank is steepened, intentionally or otherwise, lift is loss. This means we must add back pressure or be prepared to watch the altimeter unwind. Thus, the biggest problem is maintaining altitude and the wrong way to correct for it is to increase back pressure as one would in level flight. This is because in a steep turn, back pressure tends to increase the bank and that leads to loss of lift. In other words, a snowball type of situation is set up.

In order to recover from the woes from an unworthy execution of a steep turn there is but one answer. Reduce the bank with aileron control and then increase the back pressure to return to normal. The bank does not have to be rolled out more than 5° in most recoveries. Certainly 10° is the maximum. A student who goes into a lesson on steep turns with this information will impress his instructor, most assuredly.

The next hardest thing to do with a steep turn is to roll out of the turn properly. There are two reasons for this. First, the rate of turn is quite high and to roll out on a particular heading is difficult. It requires some advanced planning. The other reason that rolling out is so difficult is because we must get rid of the back pressure we have been holding or have trimmed in to aid in our execution.

Let's deal with the heading problem first. If we are using a landmark of some sort, it becomes a matter of practice. Offhand, a good reference to begin the roll out is when our ground reference passes one of the corner windscreen posts. That will begin the rollout about 30° of turn before the ground reference would pass in front of our nose. This is just a little early. That is where a little experience will pay off in later attempts. However, using an outside reference, it is easier to time the roll out.

The most accurate way to roll out of the turn is to use the heading indicator, also known as the directional gyro. The rule for doing this is to begin the roll out half the angle of bank before the desired heading. For instance, we are performing a 45° banked turn. Half the angle of bank is 22.5°. (I use 20° and it works out fine.) So if we are turning to the left, and want to roll out on a heading of 180°, we should begin the rollout when the DG passes 200°. It is as simple as that.

The back pressure held in a steep turn must be dissipated at some point. The failure to do this will result in the wings of the airplane in a level attitude and the nose climbing for the moon. Hence, it is best to begin adding forward pressure as soon as we begin the roll out. Unless one adds way too much, the nose should return to level with the wings. A few tries at this technique will net you the hand of a professional.

Now we can talk about doing the accelerated stall. In the most elementary of terms, the accelerated maneuver stall is a stall induced during a steep turn. We are not likely to induce one accidentally, as experience will show you.

As with all the stall maneuvers, this one begins with a power reduction. The speed for entry into an accelerated stall will be about 10 knots above normal takeoff speed (Fig. 5-4). This results in a speed that will let us get into the turn initially, yet is slow enough to bring the stall on rapidly as we desire. After the power has been reduced, roll directly into the 45° bank. It is important at this point that we do not allow the altitude to decline because this will give us extra airspeed and that will have to be dissipated prior to the stall.

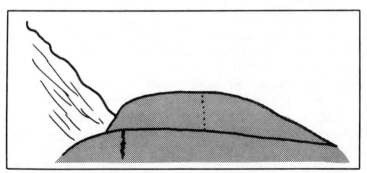

Fig. 5-4. Outside reference for an accelerated maneuver stall to the left.

Once we are in the level turn, we begin to add back pressure. Don't be afraid to really haul back on it cause that's what it's going to take. If we keep the ball centered the stall will be the easiest of the entire stall series to fly out of.

When the stall arrives the outside wing in the turn will be the one to stall—in other words, the high wing. This is due to its higher loading during the turn. As soon as it loses its lift, it will roll down to a level attitude. If we reduce the back pressure to about neutral and add power, we will fly out of the stall as if all we did was to roll out of a steep turn. It is really easy.

With the airspeed close to that used in the landing pattern it is easy to see how pulling up in a turn near to the ground could bring on a stall. It really isn't that easy to make the airplane stall; however, if the pull-up was coincidental with a strong updraft, it might just be enough. The important thing is to know how to recover if this happens inadvertently.

Well, friends and neighbors, that wraps up stalls, or the stall series as it is known. Probably no amount of reading will get one ready for doing that first power-off stall. You will most likely be apprehensive and that is normal. The guy who *isn't* somewhat apprehensive won't make a good pilot because as the old saw goes, "There are bold pilots and there are old pilots, but there are no old bold pilots."

Airspeeds and Ground Speeds

Since we have been discussing stalls and spins in the earlier pages of this chapter I thought it would be appropriate to examine some common misconceptions about flying which relate to the stall-spin syndrome in the turn from base to final. In the paragraphs below is a reprint of the FAA's Exam-O-Gram No. 17. It outlines the problem very well. Enjoy reading it. I did.

The following remarks are actual excerpts from a pilot's written report of an accident in which he was involved.

"I was climbing at an airspeed of 60 mph. I started a climbing turn to the right. The wind now became a crosswind instead of a headwind. This (lack of head wind) caused the airplane to stall—to recover from the stall I turned the airplane back into the wind . . . (later) I was in a quartering tailwind from the right . . . Went into a second stall . . . This is all I remember."

This pilot has over 100 hours, yet stalled and crashed due to an apparent misuse of controls at a slow airspeed (high angle of attack). The inspector who took this pilot's statement decided to pursue this theory with a group of student pilots. He posed this question to them.

"If the aircraft's stalling speed was 60 mph and you were flying at an airspeed of 70 mph into a 30 mph wind, what would happen if you maintained this airspeed of 70 mph, but turned downwind?" *Five of the six students said the airplane would stall.*

Is this answer correct? *No*

Is the stalling speed of an airplane a function of the airspeed or the ground speed? *The airspeed.*

Does the direction of the wind have any effect on the airspeed of an aircraft in flight? *No.*

Now to summarize our point, airspeed is the only speed which holds any significance for an airplane. Once it is off the ground, an airplane feels nothing but its own speed through the air. It makes absolutely no difference what its speed happens to be in relation to the ground. The aircraft in flight feels no wind. It simply proceeds, operating with the same mechanical efficiency, upwind, downwind, crosswind, or in no wind at all. (Note: We are referring here to a steady wind. Turbulence, gusts, or wind shears can lead to stalls even though airspeed is being maintained above the normal stalling speed. In such conditions it is wise to add a safe margin to normal climbout or approach speeds.)

I hope you got the message. The airplane will stall as a result of a high angle of attack. The high angle of attack may be generated by pulling back on the controls at cruise airspeed or it can also be generated by the airplane trying to stay aloft at a low airspeed. Turning the airplane a different direction with relation to the wind has *no* effect on airspeed.

As we mentioned earlier, wind shears such as vertical gusts can sometimes be enough to stall the airplane if the airspeed is low. There are also wind shears which occur due to temperature inversions or close proximity to weather fronts. In either case, it is possible to go from having a headwind to having a tailwind during a climb or descent. The converse is also true. Whenever we pick up an instant tailwind due to wind shear, the trouble starts. This condition will cause the airspeed to drop and that will result in a high angle of attack. Normally, the recovery from such an incident will be to lower the nose. It may surprise you how long it will take to regain the original airspeed. Hopefully, the ground will not be that close.

Chapter 6

Ground Reference Maneuvers

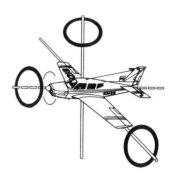

In most flying curriculums, the emphasis on ground reference maneuvers is slight indeed. Perhaps this is because more time is usually needed to learn the various takeoff and landings and to cover other basic maneuvers. The usual 35 to 40-hour training program becomes extremely tight. Because the expense of renting an airplane and instructor has always been high, ground reference maneuvers have necessarily been neglected to devote more time to the "real important" items such as stalls and landings. Nevertheless, the ground reference maneuver has a place in every flight curriculum.

What is a ground reference maneuver? The FAA defines it like this: "Ground reference maneuvers are those during which the airplane is maneuvered over a path determined by points or lines on the ground. They are of primary importance for preparing a student pilot for the maneuvering incident to landing approaches and departures after takeoff, and are essential for the safe conduct of operations involving the observation or use of ground objectives."

The two maneuvers that we will dissect in this chapter are S turns across a road and turns about a point. Both of these are relatively easy. The turn about a point is only an extention of the S turn. So beginning with the S turn maneuver, we can flow right into the turn about a point.

Ground reference maneuvers really help a student develop a feel for flying the airplane. All of the early part of the student's work is at fairly high altitudes. It is planned to be this way in order that the student can develop his technique, knowledge of maneuvers, coordination, and general feel for handling the aircraft. The upshot of all this is that the student has devoted most of his time to learning the results of control pressure on the action and attitude of the airplane. If this is allowed to continue, the student will develop a habit of fixing his attention inside the airplane. This habit can be almost impossible to break if it is allowed to go unchecked. The answer is to develop ground reference maneuvers that will draw a student's attention outside his cockpit to the ground below. All the while, he or she will have to be controlling the airplane inside as before. When these two factors team up, the result is a more proficient and safe pilot, because he can now pay attention outside the aircraft to things like traffic patterns and traffic.

To begin with, we learn ground reference maneuvers for two reasons. The first reason we discussed above. The second reason is to learn to correct the path of the airplane for wind drift. In the last chapter we discussed how the wind has no effect on the airplane flying through the air. This is true. When we compare the flight of the aircraft to the ground, however, there is a noticeable difference between the track that an airplane would take on a "no wind" day to the track it would take on a windy day.

There are several ground reference maneuvers besides the two that we will discuss in this chapter. Those explained here will be the ones that will be required on the flight check for the Private Pilot rating. There is one other ground reference maneuver we will begin with just to accent the lesson on how wind effects the track of an airplane.

Wind Drift Circles

This particular exercise takes no technique whatsoever, other than the ability to hold the aircraft in a 30° bank at 600 feet above the ground. This is why it won't be on the flight check, but I think it shows very well the direction a plane will take as it maneuvers in the wind.

All ground reference maneuvers are entered downwind. Airplanes travelling directly downwind have the highest ground speed. Therefore, the appearance of speed out the window will be at the highest. The feeling of speed will be greater than that experienced at regular flight altitudes, since ground reference maneuvers are generally done about 600 feet above the ground.

In executing this maneuver we need to pick out a ground reference. Crossroads will do fine, as will a large building. Flying directly downwind we begin our 30° banked turn abeam our reference point. We should turn toward the point and keep the bank steady. Also, it is important to be close enough to the point so that the wing will not hide it from our field of vision. As the airplane turns into the wind we should be able to notice a drop in our ground speed. It is sort of like turning upstream in a canoe—the motion seems to suspend itself. Continuing the turn, we may start to run over our chosen point of reference. The wind is pushing us towards the point. Our track will look like doodles across a note pad. As we continue to circle with 30° of bank, the point will move farther and farther away. We are, in effect, drifting downstream (Fig. 6-1).

S-Turns Across a Road

Like my old algebra teacher used to say, "Pay close attention to this one, you'll see it on the test." And you *will*, too. I never had a student take a check rider without performing this maneuver.

The object of S turns across a road is to fly a pattern of two perfect half circles of equal size on opposite sides of the road. The road should be perpendicular to the wind direction and a constant altitude should be maintained throughout the maneuver—600 feet is the norm.

Calm wind conditions rarely exist at 600 feet above the ground. This makes the S turn a good maneuver to learn to correct for wind drift. Although ground reference maneuvers can be entered from any point, downwind is the best. This allows us to use the steepest amount of bank at the outset of the maneuver.

The maneuver begins when we cross the road with our wings parallel to the road (Fig. 6-2). At that instant we should begin our

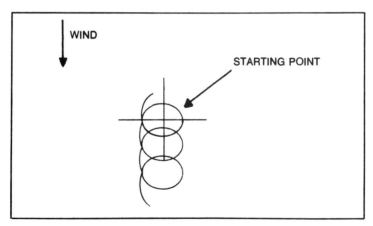

Fig. 6-1. Ground track of wind drift circles.

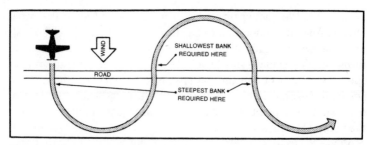

Fig. 6-2. S-turns across a road.

steepest bank. For my students, I always recommended 45°. This has the advantage of keeping the half circle rather tight and makes the student work a little harder. We only keep the bank at 45° for an instant, however. Almost immediately, we must start to reduce the bank. It is necessary to time our roll out so that the wings will be level just as we cross the road. If we level the wing prior to that, the maneuver has been done improperly. If we do not get the wings level, it is not that bad of a deal. The proper thing to do is to note the amount of bank that is left and roll to the opposite side, stopping at that same amount of bank. For example, we cross the road with 10° of bank left in a left turn. We should roll to the right immediately using 10° of bank.

As the airplane crosses the road it is facing directly into the wind. The ground speed is at its lowest. As we begin our turn to the right we begin to pick up a crosswind. This eventually becomes a tailwind and increases the ground speed. As the ground speed increases it is necessary to roll more bank into the turn. Finally, when we cross the road again we should have the maximum amount of bank equal to the original roll-in value. That value if you remember was 45°. So 45° is the necessary amount of bank needed to make the maneuver come out properly.

This maneuver is just a little demanding the first time around for nearly everyone. A little forethought, such as reading this section, can give a better edge to one's performance. The problem is that we must know what to look for to make the maneuver easier to perform.

The one thing that helps most students is to remember just how large the first semi-circle actually was. The first semi-circle ordinarily leaves a great deal to be desired. However, it will have some value to us if we notice at the 90° point in the turn how far we are from the road. Noting this, we then need to try to emulate that in the next semi-circle.

Another thing that most students try to get by with is varying the bank back and forth. In other words, they roll into the 45° bank and lessen it to 30°, then all of a sudden they believe they need 35° of bank. To steepen the bank is another way to incorrectly perform the maneuver. The way to perform this maneuver is to develop a sense of timing. It can be done in the first few attempts if the pilot knows what to watch for. What we must watch is the road. We are striving to cross the road with little or no bank. Just watch the road and slowly roll the bank out during the first semi-circle going downwind. After all, the road is our outside reference. Going into the wind we will do the opposite. We will increase the bank as needed by watching the road and timing ourselves to the crossing.

When the airplane is perpendicular to the wind during this maneuver, we are correcting for the wind in a crab. A *crab* is a drift correction angle which keeps the airplane travelling over its intended ground track. In this maneuver one would normally think of the longitudinal axis of the airplane as being tangent to the ground track pattern at each point. However, this is not the case. During the turns on either side of the road the nose will be turned towards the wind. In the beginning, downwind circle, the airplane will be crabbed towards the inside of the circle which is towards the road. In the upwind segment, where the airplane begins the turn going into the wind, the nose will crab towards the outside of the circle.

In either of the cases above, the airplane is being crabbed into the wind just as if it were maintaining a straight ground track. The amount of crab needed will depend on two things: the wind velocity, and how nearly the airplane is to a crosswind heading. The stronger the wind, the greater crab angle at any given point during the turn. Also, the nearer the airplane comes to a direct crosswind heading, the greater the crab angle will become. The maximum crab angle for either of the above reasons will come when the airplane is farthest from the road.

I suppose I should address one other problem most students have doing this exercise. That problem is maintaining altitude. It is very difficult for the first-timer both to do the maneuver very well and keep his altitude. This is one reason for doing ground reference maneuvers as we have already explained. There is really only one satisfactory method for maintaining altitude. That is to divide one's time between looking out at the road and glancing occasionally at the altimeter. One other hint that might help is to fly the airplane with only the thumb and forefinger. Since the average person tends to tense up in certain situations, it is common for the pilot, learning a new maneuver, to tighten his grip on the controls. This results very often in pulling back on the controls, causing a climb. Thus, most ground reference maneuvers usually wind up on the high side rather than on the low side.

Turns about a Point

As we mentioned earlier, there are numerous ground reference maneuvers. If you go on to a Commercial Pilot curriculum you will learn all of them. However, learning just the basics does not require that we do all of them in Private training. The turn about a point is a natural extention of the lessons learned doing S turns. When we get right down to it, the only difference between the two maneuvers is that we reverse our direction of turn with S turns. With turns about a point we just continue in the same direction, usually the left side, since that is the pilot's side and vision is greatest for him/her to that side.

To begin a turn about a point we first need to pick out a ground reference. This can be anything that is large enough to see easily from above. Crossroads work well, using the cross for the pivot point. Also, it is easy to pick references on the spokes from the pivot point. I'll explain how to use these later.

The object of the maneuver is to pivot about a point keeping a constant radius or distance from a chosen point. To maintain a constant radius will take a constantly changing bank if any wind exists. This is because variations on the radius of the turn (in reference to the air in which the airplane flies) are necessary to produce a circle of uniform radius about the object on the ground, which does not move with the wind. As with the S turn exercise, the steepest banks and highest rates of turn will be required when the airplane is flying downwind. One important item is to keep the radius long enough so that the steeper banks do not lower the wing in front of the reference.

It will be easiest to enter a turn about a point from the upwind side tracking downwind. As with S turns, this will allow the steepest bank to be used first. About 30° of bank is optimum for the entry. The aircraft is not banked until it is abeam the reference point. From that time onward the pilot's job is to watch the point and determine if and when the bank should be decreased. It should be decreased at a fairly steady rate as the nose swings into the wind. As we discussed in S turns, it is not good form to cheat and roll the bank in and out. A steady increase or decrease is desirable. If you roll too much bank out, for example, do not roll some more back in. It would be better to just hold the amount of bank that you have in already.

As the nose comes directly into the wind, the bank will be at its minimum. Some amount of bank should be left to keep the plane turning a circle. Our ground speed will be the slowest at that point, which calls for the lowest amount of bank. If you think of the second half of the scribed circle as being the same as the second half of an S turn it will be easier to relate to this maneuver. The main difference is that the direction of the turn is never reversed in a turn about a point.

Now, here are some tips for executing the maneuver properly. To begin with, divide the maneuver into four obvious parts. There is the downwind entry, and following that is the crosswind point on the downwind half of the cirle. Next, comes the point where the airplane is headed directly into the wind which we will call the upwind point. Ground speed is the slowest here. Following this point is the crosswind point on the upwind side of the circle. If we use a crossroads for our reference these points are readily identifiable by bushes, culverts, driveways, buildings, etc.

The idea is to use these point as aiming points. We know, after a very short time doing this maneuver, where we are supposed to be increasing bank and where we should be decreasing bank. As long as we are doing that we can make the rest of the job a snap. Once we enter the maneuver we should turn our attention away from the pivot reference in the center *occasionally* towards these aiming points. These points should be picked about an equal distance from the pivot as the original downwind entry point. Once this has been done we can reduce our job to flying an arc between these aiming points. Of course, we should be steadily increasing or decreasing the bank as our relation to the wind demands.

Another tip is to use the wingtip. There is an area radiating out from the wingtip of about 15° either side of the leading edge and trailing edge of the wing. For the most part, if the reference point at the center of the circle is within these boundaries the pilot is doing the maneuver satisfactorily. If the point moves behind the wing and out of these bounds then the pilot should increase the angle of bank. By the same token, if the point moves in front of the boundary the bank is too steep and needs to be decreased. These are handy little rules of thumb. If you can remember them you will improve your overall flying performance and impress the instructor and later the examiner.

As with the S turn, altitude maintenance is another problem. However, by the time a student graduates to this maneuver it is ordinarily easier to divide attention between outside references and inside references.

There are several other ground reference maneuvers that the Private Pilot student may be asked to perform by an instructor. With all of these various maneuvers the principle remains the same. When the ground speed is high, a larger angle of bank will be needed to hold the aircraft over the desired ground track.

One other basic fundamental—if the direction of the crab is hard to envision in a turn try the following: Find a straight road or railroad track to use for a ground reference. Chances are that the wind will be a crosswind. Line up the longitudinal axis of the airplane with the track or road. Soon the wind will push the airplane off to one side or the other. Now try to stay right directly over the road. The nose will have to be angled to the side that the road drifted towards in the first part of this experience. Finally, a crab angle will be apparent in order to remain over the road or track. Next, turn around and head back the way you just came (a 180° turn). Which side is the wind from? In relation to the plane, the wind is on the other side. In relation to the reference on the ground the wind is from the same side, only now the nose is pointed in the opposite direction.

How does all this relate to the curved paths that we have been discussing? The relationship is in the fact that the ground reference maneuvers change the direction of the plane. This changes the side that the wind acts upon. It also changes the amount of bank we need to fly the correct ground track. Thus, if at any moment we can relate our position in the maneuver as to being on a straight line course we can more accurately deal with the wind.

So, there it is. The biggest help to performing ground reference maneuvers properly is being aware of the wind. Watch the water on lakes or how tall grass is being blown to find the wind direction. Then throughout the maneuvers check something that can verify the wind direction. It makes the maneuvers so much easier.

Chapter 7

Special Takeoffs

In Chapter 2 we talked about that all-important first flight. The flight itself began with a normal takeoff. Granted, a student's first takeoff usually is not *that* normal, but at least understand that is what we were striving for. So, what is a normal takeoff? A normal takeoff is one in which the nose is rotated at a time in which the plane will actually fly itself off the ground. No special flap settings are used and no special manipulation of the controls is needed.

If the above definition is correct, then special takeoffs include either special flaps or control manipulations. In some cases we may need to use all sorts of combinations.

The Crosswind Takeoff

Before we get started into the particulars of doing a crosswind takeoff one point should be extremely clear. A crosswind takeoff is a technique that we use *any* time there is a crosswind. In the succeeding pages we will discuss short field takeoffs and soft field takeoffs. Both of these takeoffs should be made with crosswind takeoff procedures if a crosswind exists. Just because we are making a soft field takeoff does not mean we cannot use crosswind procedures. Thus, where a soft and short field takeoff are separate procedures that might include a crosswind technique, the crosswind takeoff may be done at any time. Nevertheless, the technique for taking an airplane off in a crosswind does require some knowledge and practice—a great deal of practice. In fact, nine out of ten private pilots cannot make a satisfactory crosswind takeoff six months after they get their rating. The reason is that they don't think it is

117

important enough to bother with. Let me tell you right now, you may get away with that for years, but *someday* . . .

Honestly, there is one other reason why pilots do not make effective crosswind takeoffs. This reason is due to the limpness of the controls at the beginning of the takeoff roll. They don't "feel" like they are doing anything to help. But they are doing more and more as airspeed builds.

The crosswind takeoff is needed any time the wind is not blowing directly down the runway. The reason the technique was developed was to keep the airplane traveling a line between the runway lights. Without some sort of control correction for the wind, the airplane will weathervane into the wind. If the wind is coming from a 60° angle to the runway, eventually, the airplane will head in that direction and net result is a ride through the boondocks. This will remove things from the airplane like landing gear and pilot during the sudden stop.

The technique for crosswind takeoffs is really quite simple. Let us pretend that there is a crosswind from the right side of the runway. The tendency of the wind is to get under that right wing and push it up. That would result in the right main gear possibly becoming airborne. When that happens, the inherent stability of the tricycle gear is upset. This could lead to the left wingtip dragging the ground. The end result could be simply a scratched wing or a full blown cartwheel of the aircraft. To guard against this happening we use a technique that is used during taxiing the aircraft. In regards to using the ailerons for crosswind protection the following should be done: If the wind is in front of the airplane turn the ailerons into the wind. If the wind is behind the airplane turn the ailerons away from the wind.

When this rule is followed closely the ailerons are positioned in such a way to reduce lift to the maximum possible extent on the windy side of the aircraft. Therefore, the chance of the wing wanting to fly on that side is less, and it will tend to love the ground.

At this point we need to apply this rule to a takeoff. This is done by rolling-out on the runway and rolling the ailerons *completely* into the wind. Now, any idiot knows that if the airplane gets going fast enough with the ailerons fully deflected the wings are going to want to bank. If we do this on the runway we will scratch a wing or worse on the very side we are trying to protect from the wind. So we must roll the aileron correction out slowly as the aircraft gains speed. The amount that we roll-out will be governed by the feel of the controls. What we intend to do is to keep a gentle pressure on the ailerons. When the pressure seems too stiff, or we can feel the other wing lifting, we need to decrease the pressure, but only slightly. To release the pressure on the ailerons now would only cause definite control problems for the aircraft.

Close to the end of the takeoff roll we can expect that the wind will still be blowing. Therefore, we can also expect that we will need at least a little pressure towards the crosswind. At the moment of liftoff, we will still have just a little aileron correction left. The wing will dip as the wheels leave the ground. If they don't, the procedure has not been done properly. Next, we can expect the airplane to weathervane into the wind. That is, if we leave the crosswind control input alone it will automatically set up its own proper crab angle. This is what we want because the airplane will then track down the extended centerline of the runway until we are ready to begin our crosswind turn.

In summary, we begin the takeoff roll with full aileron correction. As airspeed builds throughout the takeoff roll, we gradually decrease the amount of correction that we are holding against the crosswind. At the moment of liftoff the ailerons have a little bit of correction left and the wing should dip slightly. At that point we should neutralize the ailerons and let the airplane weathervane into the wind. This sets up an automatic crab angle which we should endeavor to keep until we begin our crosswind turn in the traffic pattern.

It all sounds so very simple. Yet it is amazing how many rated pilots don't fool with the technique at all. As a result, they endanger lives and property. The very least they do is wear out tires early.

The Short Field Takeoff

If you have read any other aviation publications prior to this one on learning to fly, you may have come across the ubiquitous 50-foot obstacle. To read some literature, one would think that there must be a 50-foot wall at the end of at least half of the runways in the country. Luckily, there are very few obstacles to be cleared at most airports. The trouble is with those little country grass strips out where the fishing is so good. We can use these strips for their intended purpose but we usually need a few tricks of the trade to make it a safe venture. In order to miss the pine trees at the end of that 2000-foot strip we need to get airborne in a hurry. That is what the short field takeoff is all about.

For the short field takeoff we must go a long way back to the first chapter or so. Remember we discussed the difference between best *angle* of climb and the best *rate* of climb speed? Which of those speeds do we like to use for a short field takeoff? If you said V_x (as it is known) for the best angle of climb speed you were correct. The best angle of climb speed will give us the most altitude gain for the least distance traveled forward by the airplane. In simpler terms, if we use the best angle of climb speed instead of some other speed, by the time we cross the departure end of the runway we will be the highest we can possibly be. Therein lies the reason for using this speed, because the trees are always at the end and never in the middle of the runway.

The actual performance of a short field takeoff will often surprise the student trying it for the first time. Because the deck angle is so high compared to a normal takeoff, students normally want to lower the nose for fear they will stall the airplane. I'll admit it takes precise airspeed control to do this type of takeoff right. However, we don't have to keep the airplane at that speed for very long. Also, the best angle of climb speed is not that close to stall speed.

Let's go through the technique. As with all takeoffs we need to take the runway and get lined up with center. In actual short field situations we should pull as close to the threshold of the runway as we possibly can. There is nothing so useless as runway behind and altitude above. After doing this we should stand on the brakes and run the engine up to maximum power. There has been, in years gone by, arguments on whether this is the best procedure or not. I believe it is because it allows the engine to attain its max power prior to covering any ground. Upon release of the brakes, the airplane will begin its roll down the runway. Since the best angle of climb speed will be higher than normal takeoff speed we must endeavor to keep the nosewheel on the ground until that speed is reached. There is a slight danger of the main wheels coming off the ground prior to the nose wheel, which is a condition called wheelbarrowing. If only slight pressure is applied forward on the yoke, this won't happen. Too much pressure might cause it to wheelbarrow.

The airspeed must be monitored very closely and the takeoff roll should be as straight as possible. This means we might have to use crosswind techniques as well. A wandering takeoff roll will prolong the moment of liftoff and with trees or even worse, a legendary 50-foot obstacle to clear, we need to get off as early as possible. Upon reaching the chosen speed for the aircraft we are using, virtually pop the airplane off the ground. Pitch the nose up very strongly and keep the airspeed right on V_x. Fifty feet will fall away extremely fast. Once we have cleared the obstacle we can begin to lower the nose and gain airspeed, then return to normal climb speed.

As anyone can see, the technique seems simple enough. The hardest thing to overcome is the tendency to want to put the nose down prior to reaching the altitude necessary to clear the obstacle. About all that it takes is faith that what the instructor has said is correct, and determination that you will do it that way no matter what. Of course, accurate airspeed maintenance is not that easy any way you look at it. It takes practice.

The technique is made a little more difficult whenever there is a crosswind. The steering on the nosewheel (or directional control with tailwheel airplanes) gets a little squirrely. This problem stems from having to keep the airplane on the ground beyond the normal takeoff speed. There is no way to help this out except to use extra determination and keep the wheels on the ground until reaching V_x.

The Soft Field Takeoff

If you live where it might rain often you will need this technique in your bag of tricks. Even the northern states have the problem when the frost goes out of the ground. The small diameter wheels that are on airplanes are good for getting stuck. Some bush pilots add larger tires to their airplanes to aid in getting out of muddy places. That might not be a bad idea if you have to get in the mud every day. For the very occasional encounters with a muddy field a soft field technique will suffice. If the field is *too* muddy, forget flying altogether.

The soft field takeoff is a technique that pilots use to get the airplane airborne with the least amount of resistance and in the least amount of runway. We decrease the resistance of the mud by holding the nosewheel out of the mud. To shorten the takeoff roll we use flaps.

The amount of flaps that is necessary is obvious once several things are considered. Namely, we want to reduce resistance in the mud and in the air. Too large an increment of flaps will develop too much drag and that would increase the length of our takeoff roll. Yet, we know we should use some flaps because that will enable the airplane to fly at a slower airspeed and thus become airborne at an earlier point on the runway. The flap increment most chosen in light aircraft for this procedure is 10 degrees, which gives the airplane added lift without the ponderous drag that come with flap settings greater than 20 degrees. We could, I suppose, use 20 degrees of flaps for the procedure. In that case, though, the climb performance of the airplane would suffer somewhat. If there were any type of obstacle to be cleared, we might be at a disadvantage. Unfortunately, too many times soft fields are also short fields. In that case we must weigh the variables and choose a compromise solution.

In preparation for the check ride the student will only be required to perform a truly soft field takeoff. In light of that, that will be all we examine here.

A soft field takeoff may begin from the moment the airplane begins to taxi from the tie-down. If the taxiway is muddy we need to begin our technique right away. In such a case, we pull the flight controls back into our lap. Doing this takes as much weight off the nose wheel as possible. We are decreasing the resistance that the mud and soft ground has on the airplane.

In actual soft field conditions it makes sense to do the run-up on the way to the runway instead of stopping. To stop may mean that we may need outside help to start again or that we might damage the airplane trying to get it moving again. However, in training we will stop and do the run-up at the normal time and place.

From the run-up area we will lower about 10° of flaps. As we begin the taxi from the run-up area to the runway we should pull the yoke back into our lap. As we turn on to the runway we should feed in full power and continue holding the controls all the way back. As soon as the nose wheel comes off the ground it will become necessary to release some of the back pressure. Holding too much back pressure is as bad as not enough; we might drag the tail. Another reason is that form drag is greatly increased and will slow our time to achieving V_x. It is very important, nonetheless, to keep the nose wheel off the ground as friction with the ground is lessened.

At some point the stall warning will start to sing. Let it sing. Let the airplane stumble into the air riding on ground effect. Ground effect is only about a wingspan deep beginning at the ground. To allow the plane to fly out of this margin of safety is to invite a stall. If we stay within this wingspan of distance, however, the airplane will not stall if it has already pulled itself into the air. The thing we must do as soon as the main wheels break ground is to lower the nose. It is common to lower the nose too much and find the airplane landing again. The idea is to level off about five feet above the runway and fly along until normal climb speed is attained. Once we have achieved this speed we begin our climb.

After any possible obstacles are cleared and the altitude is 100 feet or so above the ground it is safe to think about retracting the flaps. Many pilots that I have ridden with just pop all the flaps up at once. Personally, I think this is bad technique. The airplane sags as it struggles to climb. There is a possibility of creating a stall if the airspeed is not high enough. A better technique is to raise the flaps very slowly or in increments. This minimizes the chances for creating a stall and feels a great deal better to passengers who are not as used to flying as we are.

It seems easy enough to do, yet many students have a problem learning this technique. A woman I was teaching froze up on me one morning and held the controls all the way back from the beginning of the takeoff roll until the airplane broke ground. In doing this she got the airplane to stall even in ground effect. With the crosswind we had that morning we soon drifted off the side of the runway. The left wing started to dip into the grass as I used both hands on the controls to overcome her death grip on the controls. I thought the airplane was about to cartwheel beside the runway and roll up in a wad somewhere. Luck was with us though, as ground effect got the airplane flying in a jiffy. Guiding the airplane back over the runway we landed and then I went inside for some coffee.

To summarize the technique: hold back pressure all the way back until the nose wheel comes off the ground. In ground effect, fly the airplane off at the slowest possible speed. Once airborne, concentrate on picking up airspeed until it reaches the normal climb value. Once you have climb airspeed, climb above obstacles at hand and milk the flaps up. From there on the process is strictly a normal climb to altitude.

Chapter 8

Special Landing Techniques

Earlier in the book we talked about the first lesson. In that chapter we went through how the traffic pattern is flown. We will discuss it again with a few photos to help you visualize how to do it properly.

Basically, there are four sides to the traffic pattern just as there are four sides to every rectangle. If one of the long sides is represented by the runway then all others must be flown in the air. After takeoff we make a right angle turn (either left or right, it doesn't matter). Once we have established ourselves in that direction we are flying the crosswind leg. This particular leg is the only one that can be flown during takeoff as well as prior to landing. The other two sides are the downwind leg and the base leg. The final leg is the same as the runway leg.

In entering the traffic pattern for landing, we may do it several ways (Fig. 8-1). At an airport governed by a control tower, we do it any way they ask us to do it. At an uncontrolled airport it is best to fly the entire traffic pattern because that is what everyone is expecting the other guy to do. A 45° diagonal entry to the downwind leg is the formal way of entering a traffic pattern. Another method that is just as good and generally used to cross the airport from one side to the side of the active traffic pattern is known as entering crosswind. In using this method we fly at right angles across and above the runway until turning downwind.

The power is reduced at the point that is abeam the threshold. At this point the descent to the runway begins. We trim the airplane for the descent and sometimes add our first increment of flaps (Fig. 8-2).

Fig. 8-1. Proper spacing between the runway and the plane is important to flying a good pattern. In this C-150, the runway should appear to be halfway up the strut.

The base leg follows downwind and is where we turn towards the runway to close the end of the rectangle. At this point we reach the "key position." The key position is shown in Figs. 8-3 and 8-4. It is at this point that we adjust our rate of descent to the runway. At the key position it is possible to judge whether the pattern we are flying is going to let us land short of the runway or not. This will take a little experience to master, but you should know this beforehand.

Fig. 8-2. When we are opposite the landing threshold we retard the power and trim the airplane for the descent to the runway.

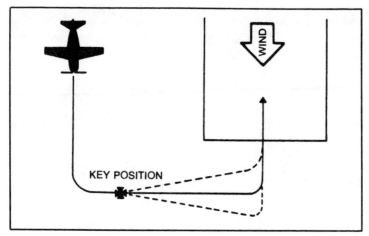

Fig. 8-3. This illustration shows the exact location of the "key position." From here we can adjust the pattern to correlate with the altitude that we have left.

If we are low on the approach we add power. If we are high on the approach we reduce power (if the power is already at idle, then we add more flaps to steepen the approach).

Once we have turned final approach we should carefully line up with the center of the runway (Fig. 8-5). Many students are satisfied if the runway is just anywhere in front of the airplane. Most instructors are not. Believe me, a student pilot needs to be in the middle of the runway for safety's sake because the airplane in the hands of a

Fig. 8-4. This is what key position looks like in right-hand traffic.

126

novice might do just about anything. Final approach is also the time where we make final adjustments in the descent and correct for any crosswind that exists. We can correct for descent by the same method as before, with power or additional flaps. In the event that we are very high on final approach, we can slip the airplane for quick altitude loss. In the following paragraphs we will discuss how to do both kinds of slips.

Forward Slips and Sideslips

As you noticed in the subheading above there are in fact two kinds of slips (Fig. 8-6). They are very different and are used for very different types of situations. Let me explain further. The forward slip is used to increase airplane drag and thus increase the rate of descent without increasing airspeed dramatically. In other words, we can make an aircraft fly an airspeed of 80 knots and descend at 2000 to 2500 feet per minute. This would be impossible by just putting the nose down or lowering flaps.

The sideslip, on the other hand, is used to control the direction of the aircraft during landing in crosswinds. Though both slips use crossed controls, we tend not to use as much rudder in the average sideslip. Now let us go through both techniques.

The average altitude for most light aircraft that are on a mile final is about 400 feet. Sometimes, though, our timing just isn't what it should be or there may even be a tailwind above the surface winds (known as *wind shear*). In either case we might wind up on very short final with excessive altitude. Our choice is to go around and set up for another landing (which, by the way, is not a bad choice), or we may slip the airplane for quick loss of altitude.

Fig. 8-5. On final approach there is only one way to stretch the glide—add power!

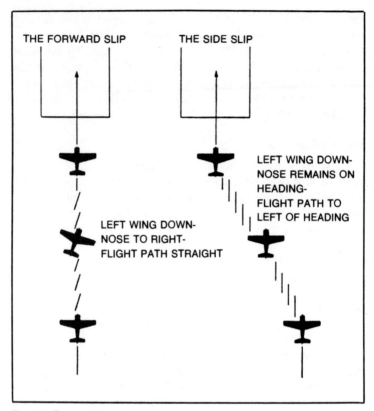

THE FORWARD SLIP

THE SIDE SLIP

LEFT WING DOWN-
NOSE REMAINS ON
HEADING-
FLIGHT PATH TO
LEFT OF HEADING

LEFT WING DOWN-
NOSE TO RIGHT-
FLIGHT PATH STRAIGHT

Fig. 8-6. Forward slip and sideslip.

To get the aircraft into a forward slip is really quite simple. Beware, however—it is uncomfortable to most passengers. All we need to do is to push either rudder to the floor. This does two things. First, it puts the side of the airplane into the slipstream and that creates extra drag. The other thing this does is to cause the airplane to skid into a turn. To compensate for the skid and an unwanted change in direction we must add opposite aileron. In other words, if the rudder is making us turn to the left, we need to turn the ailerons to the right. Almost a full deflection will be needed if the rudder is deflected fully. Also, please understand that now the controls are *crossed*. They are working *against* each other. It now becomes imperative that we lower the nose to keep the airspeed within allowable limits. If we raise the nose it is easy to run into a high angle of attack and stall. Stalling with the controls crossed will result in an immediate spin. This is no exaggeration, the airplane *will* spin with crossed controls. This is not to say that the slip is a dangerous maneuver. It is very commonly done, only we should keep the nose below the horizon at all times.

Another consideration in doing forward slips is the accuracy of the airspeed indicator. Turning the airplane sideways, so to speak, affects the angle at which the pitot tube and static port receive outside air. In one direction, the airspeed will read higher true. In the other it will read lower than the true airspeed. Since all airplanes are set up just a little different, ask your instructor which side does what on your airplane.

A high rate of descent is what the forward slip will give us. As a result, we must be on our toes to recover and resume a normal glide path to the runway. It is always favorable to recover a little early until one is totally familiar with the technique. Since the rate of descent will be close to 2000 feet per minute, an early recovery can be desirable. To begin the recovery all we must do is to neutralize the controls. The plane will resume its normal flight characteristics at once.

In summary, the forward slip puts the side of the airplane into the slipstream. With this extra area it becomes impossible for the wing to counteract this drag with lift, especially when the engine is at idle power. (Incidentally, it is foolish to want to lose altitude quickly and not retard the throttle completely.) Once we have induced the slip, airspeed indication may be erroneous, so keep the nose down. Prepare to recover from the slip early until familiar with the technique and then land normally.

Another condition in which the forward slip is invaluable is in an emergency descent. For example, if there were a fire or smoke of unknown origin we might need to get down in a hurry. The forward slip would be the answer. If the fire was under the cowling, it might be on one side or the other. We could slip the flames to the side of the aircraft that was not into the wind. This would do two things. It would keep the flames from being fed as much air, and keep them away from the cockpit. So the forward slip is not necessary only for landing.

The sideslip, on the other hand, is for landing only. We use the sideslip for controlling the aircraft in a crosswind. As we mentioned in the last chapter, the airplane tends to crab into the wind for a given ground track when it is airborne. During landing, we often approach the runway in a crab because of the crosswind. If we continued in that fashion we would eventually touch down in a crab on the runway. This is undesirable due to the damage that can be inflicted on the landing gear, not to mention the chance of flipping the aircraft over.

It is only fair to say that there are two basic landing techniques. One is where the pilot will use the sideslip for the last 100 feet of descent or so and land with the nose wheel centered with the plane going straight ahead. The other method, one in which you will see more often with large jet aircraft, is to keep the airplane in a crab until an instant before touchdown. At that time the airplane is swung with the rudder to line up with the centerline of the runway and virtually plopped down. The former method is the one we will discuss. The other method you can perfect on your own.

On final approach to the runway it is normal for us to maintain a crab angle. This crab enables us to track inbound on the extended centerline of the runway. At some point though, we must straighten the airplane in order to land with the wheels lined up to the direction of motion. Hence, the sideslip was invented. The sideslip is so named because it actually slips side ways to the landing. The point at which we transition from the crab angle to the sideslip is a matter of personal preference. A good altitude is about 100 feet. That gives the airplane plenty of time to swing around and get lined up.

Let's pretend we have a sturdy crosswind from the right side of the runway. In the crab, the nose of the airplane will be pointed to the right. In order to land we must swing the nose around to match the center line. This would mean that we need to push the left rudder to bring it in line with the center of the runway. If we use the rudder alone, the wind will eventually push the airplane to the left side of the runway or farther. The only other controls we have to prevent this are the ailerons. As with crosswind taxiing and takeoff, we must turn the ailerons into the wind. In the air this will cause the wing to be lowered. In essence, the lowering of the wing begins a turn in the direction of the crosswind. The rudder keeps the airplane from turning that direction. The net result is that the airplane tracks straight toward the center of the runway.

The actual touchdown in a crosswind landing is almost the reverse of a crosswind takeoff. To begin with, the upwind wheel will touch the pavement first. As airspeed falls off, the downwind wheel will be lowered to the pavement, then finally the nosewheel. While all this is happening, the ailerons are slowly being turned farther and farther into the wind. At the end of the rollout the ailerons should be completely turned into the wind.

A crosswind landing can be broken into a 1-2-3 method like this:

1. Step on the rudder opposite the wind and swing the nose around to line up with the runway.
2. Add enough aileron to keep the plane from drifting to the left or right as the wind dictates.
3. Touch down with the upwind wheel first.
4. Slowly roll the remaining aileron travel into the wind to protect the wing from being lifted by the crosswind.

Soft Field Landings

As with the soft field takeoff, there may be a time when we have to land in the slop. Probably one of the most demanding of all techniques is the soft field landing. The object of this landing is to touch down on the runway as softly as possible. The speed at the time of the flare for landing is just above the stall. If we flare too much, we might stall the airplane and really squat one on. Of course, in an actual soft field doing something like that might stick the wheels in the mud and bring us to a sudden stop. That wouldn't be good for man or plane.

The traffic pattern for this landing is standard. (All traffic patterns are the same for all landings. Only the final approach is adjusted to give a different result.) As with the short field and soft field takeoffs, crosswind technique may be necessary to successfully complete either a soft field or short field landing.

Soft field landings can be done with any amount of flaps. An intermediate flap setting such as 20° is the best, however. The reasoning behind this is that the wing will provide extra lift at a lower airspeed. In turn, the ground speed at touchdown will be as low as we can possibly get it. This is what we are striving for. If we use full flaps for a soft field landing, we won't lower our touchdown speed very much and we run the risk of dropping the airplane in. That's a definite no-no if the landing is to be soft.

The technique for a soft field landing is normal to most landings with the exception of the intermediate flap setting. Also, where floating the aircraft in the flare is not generally desirable, it is with this type of landing. On final approach the glide to touchdown will be very close to that of an airplane using no flaps at all. We should be shooting for an approach speed about 1.3 times the stalling speed of the aircraft with 20° of flaps. For example, a good speed for a Cessna 150 is about 70 mph. This speed needs to be held as long as possible especially during the beginning stages of the flare.

The airspeed is critical on final approach as is the way the flare is done before touchdown. It takes a good steady hand to work the flare to its max. What we are trying to do during the flare is bleed the airspeed off without letting the wheels touch prematurely. We are also attempting to keep the nosewheel high as in a soft field takeoff. When we touch down, the airplane should be only a knot or so above the stall and carrying its nosewheel quite high. When the mains touch we can keep as much weight as possible off the nosewheel until we arrive on firmer ground. Believe me, it will take a great deal of practice to perfect this type of landing.

In summary, the soft field landing is very much a normal approach. It does require a little more precise airspeed management than the average landing approach. We use an intermediate flap setting for landing and touchdown just above the stall. During the flare we try to float the aircraft as long as possible to result in a low touchdown speed and a nose-high attitude. The nosewheel is carried high to result in the least amount of resistance after touchdown.

The Short Field Landing

The short field landing takes a great deal of planning. Even on downwind the pilot must pick the place he intends to land. Of course, no short field landing would be complete without the omnipresent 50-foot obstacle. In picking our place of touchdown we must gauge how far our trajectory will take us beyond the 50-foot obstacle. A certain amount of experience will help in this category. The landing is even more difficult due to the fact that we cannot actually see the 50-foot monolith.

All short field approaches are consummated with full flaps. Full flaps will give us the low airspeeds required to give us low touchdown speeds. Also, with full flaps an airplane tends to be through flying once the wheels have touched the ground. Since we intend to touch down firmly and get stopped as quickly as possible, this is advantageous.

As with the soft field approach, airspeed maintenance needs to be accurate. We will be operating at 1.3 V_{so} or 1.3 the stalling speed of the airplane with the flaps fully extended. It is necessary to fly as close to this speed as possible because it is the slowest safe speed to fly. To fly faster than this speed will extend our flare and rollout and we will not stop in as short a distance as desired.

The short field landing is not much different from a normal landing until we turn on the final approach. At this point we must gauge our 50-foot obstacle because it controls our glide path (Fig. 8-7). We will be controlling our glide path down final approach with corrections from the throttle. For most of the approach we will be busy watching two flight instruments. The airspeed indicator and the altimeter will take a steady scan. We know why to watch the airspeed indicator, but why the altimeter? Because we have to cross the abominable 50-foot obstacle with a slight margin of safety, and we must meter our descent to barely clear the obstacle.

Landing short provides the student with a great temptation to cross the obstacle and then dive for the runway in an effort to get down as short as possible. This is usually self-defeating. Any time we lower the nose we will by the nature of gravity gain speed. It doesn't matter for how limited a time we dive, the speed will increase. In the end, when we flare, we must dissipate this little bit of excess speed. To do that we will use extra runway while the speed is lost. Therefore, the proper way to cross the obstacle is in a stabilized approach. The airplane will be slowed to $1.3\,V_{so}$ and when we cross the obstacle our speed should be just that. As the flare begins, our speed should still be at $1.3\,V_{so}$ and this will minimize the length of runway used.

Upon touchdown, the nosewheel is lowered quickly to the ground. At that time we must apply maximum braking. If everything has fallen into place the landing will be as short as possible. In an actual short field landing, the amount of braking needed will be obvious. For practice and testing purposes though, we should use maximum braking or just call to the instructor that max braking is being simulated.

Fig. 8-7. This photo shows key position from left hand traffic. From this position we must judge our trajectory that will cross the ubiquitous 50-foot obstacle.

In summary, the short field landing is a stabilized approach. The airspeed and altimeter must be monitored closely. After clearing the obstacle, the nose should not be lowered because of the gain in airspeed that would still have to be dissipated. Upon touchdown, maximum braking is applied to bring the aircraft to a stop in the least amount of distance. Crosswind correction may need to be applied.

The Downwind Turn

A great deal has been written and said about the downwind turn. A few think it is the reason so many stall-spin accidents happen on the turn from base to final or from downwind to base. Nothing could be further from the truth. The downwind turn is a myth as far as being the cause for so many accidents.

A downwind turn is a turn from a downwind heading to a heading more into the wind. When a plane is maneuvering close to the ground, as in the traffic pattern, the speed of the aircraft is more apparent. We already know that ground speed is the greatest tracking downwind.

In brisk winds, something does happen in downwind turns. If the pilot is aware of this speed he may feel he is overrunning his desired pattern. In other words, it is being stretched out more than he would like. To correct for this, the pilot tightens the turn which increases the load factor. Next, if the pattern has been stretched out he may try to stretch his glide. How does he go about it? Well, the proper way is to add power and be patient. The pilot who gets himself in trouble goes about it another way. He raises the nose and begins to trade airspeed for altitude. When this is coupled with the steepened bank, the chance for a stall is imminent. In many cases it has happened, catching the pilot unprepared and unaware. So you see it is not the downwind turn. It is the increased load factor and higher angle of attack. Be careful about this one point, because it is the most common cause of accidents other than continuing VFR into adverse weather conditions.

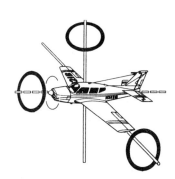

Index

Edited by Steven Mesner